I0043711

William Stirling

Outlines of Practical Physiology

William Stirling

Outlines of Practical Physiology

ISBN/EAN: 9783744718882

Printed in Europe, USA, Canada, Australia, Japan

Cover: Foto ©berggeist007 / pixelio.de

More available books at **www.hansebooks.com**

OUTLINES

OF

PRACTICAL PHYSIOLOGY:

BEING A

Manual for the Physiological Laboratory,

INCLUDING

CHEMICAL AND EXPERIMENTAL PHYSIOLOGY, WITH REFERENCE TO PRACTICAL MEDICINE.

BY

WILLIAM STIRLING, M.D., Sc.D.,

PROFESSOR IN THE VICTORIA UNIVERSITY, BRACKENBURY PROFESSOR OF PHYSIOLOGY AND HISTOLOGY IN THE OWENS COLLEGE, MANCHESTER, AND EXAMINER IN PHYSIOLOGY IN THE UNIVERSITY OF OXFORD.

With 142 Illustrations.

CHARLES GRIFFIN & CO., LONDON
PHILADELPHIA:
P. BLAKISTON, SON & CO.,
No. 1012 WALNUT STREET.
1888.

Dedicated

TO

MY REVERED AND BELOVED MASTER,

CARL LUDWIG.

PREFACE.

THE present work was written to supply the wants of the Students attending the course of "Practical Physiology" in The Owens College, but it is hoped that it will be found useful also to students of Medicine and Science in other Colleges and Universities.

This course was instituted by my predecessor, Dr. Arthur Gamgee, and extended by him and the late Mr. W. H. Waters, M.A. Mr. Waters himself intended to write a short Manual for the guidance of the members of this class, but he was struck down, to the sincere regret of many of us, before he could accomplish his purpose. The experiments herein described are performed by every member of the class, and they are practically a repetition of some of those which I am in the habit of showing to illustrate my lectures on Physiology. It will be seen that much of the apparatus is of a simple character. Of course, no experiments are given which involve the infliction of pain upon living animals.

A considerable portion of the chemical part was in print several years ago, although in a somewhat different form. In the preparation of the experimental part, however, I have had the advantage of knowing the system and methods that were followed by Dr. Gamgee and Mr. Waters.

In arranging the experiments, the following works have afforded me much valuable information :—Dr. A. Gamgee's *Physiological Chemistry, The Handbook for the Physiological Laboratory, Practical Exercises on Physiology* by Professor Burdon-Sanderson, *Practical Physiology* by Professor Foster and Mr. Langley, Professor Grainger Stewart's *Lectures on*

Albuminuria, the printed slips of Dr. Sheridan Lea, Gruen-hagen's *Lehrbuch der Physiologie*, Helmholtz's *Physiologische Optik*, Krukenberg's *Grundriss der Medicinisch-Chemischen Analyse*, and Hoppe-Seyler's *Handbuch der chemischen Analyse*.

One feature valuable in any Text-book designed for practical ends, viz., the keeping in view of the fact that "the Student of to-day becomes the Practitioner of to-morrow," has been constantly before me. Hence, the aim has been so to arrange the exercises as to give them a bearing on, and to lead gradually up to, the methods used in Practical Medicine.

I am indebted to Mr. John T. Millett, B.Sc., and to my demonstrator, Mr. A. F. S. Kent, B.A., for reading the proof-sheets.

Many of the illustrations were drawn for me by my pupil, Mr. Philip Worley, and some were photographed from apparatus in the Physiological Laboratory of The Owens College, by Mr. William Charles, Steward of this department. I have also to express my thanks to several scientific instrument-makers, including Messrs. Browning, Elliott, Maw Son & Thompson, Jung of Heidelberg, Rothe of Prague, and Carl Reichert of Vienna. Some of the illustrations are from the *Text-Book of Physiology* by Landois and Stirling.

WM. STIRLING.

THE OWENS COLLEGE,
MANCHESTER, *January*, 1888.

CONTENTS.

LIST OF ILLUSTRATIONS.

PART I.—CHEMICAL PHYSIOLOGY.

LESSON I.

THE PROTEIDS AND ALBUMENOIDS.

The White of Egg may be taken as the Type.

1. Preparation of a Solution.—Place the white of an egg in a porcelain capsule (taking care that none of the yolk escapes), and cut it freely many times with scissors to disintegrate the membranes, and thus liberate the albumin. Add twenty volumes of distilled water, and place the mixture in a flask. Shake well until it froths freely. Cork the flask, and invert it, mouth downwards, over a porcelain capsule ; the membranes will rise to the surface, and, after a time, if the cork be gently withdrawn to allow the fluid to escape, a comparatively clear, or slightly opalescent, fluid will be obtained. If the fluid be too opalescent, strain through flannel or several folds of muslin. Such a solution filters slowly, so that it is better to employ several small filters if a clearer solution be required. If the fluid be alkaline, neutralise it. This solution contains about 5 per cent. of albumin, and diffuses slowly through animal membranes.

(*a.*) To some of the fluid in a test-tube add strong nitric acid = a precipitate, which on being boiled turns yellow. Allow the liquid to cool, and add strong ammonia = an orange precipitate or colour (**Xanthoproteic Reaction**).

(*b.*) To another portion add Millon's Reagent = a precipitate which becomes reddish on boiling. A red colour of the fluid is obtained if only a trace of proteid be present.

1

(c.) To a third portion add a drop or two of *very dilute* solution of cupric sulphate, and then a solution of caustic soda (or potash) = a *violet* colour (Biuret Reaction).

(d.) Make a fourth portion strongly acid with acetic acid, and add potassic ferrocyanide = a white precipitate.

(e.) Heat a portion of the neutral solution = a coagulum about 70° C.

(f.) To a solution of white of egg add glacial acetic acid, and heat to get it in solution; gradually add concentrated sulphuric acid = a violet colour (**The Reaction of Adamkiewicz**).

(g.) Wash finely powdered albumin first with alcohol and then with cold ether, and heat the washed residue with concentrated hydrochloric acid = a deep violet-blue colour. This is best done in a white porcelain capsule, or on a filter-paper in a funnel; in the latter case, the boiling acid is poured gently down the side of the filter-paper (**Liebermann's Reaction**).

(h.) **Non-diffusibility of Proteids.**—Place some of the solution either in a dialyser, or in a sausage-paper made of parchment-paper, and suspend the latter by means of a glass rod thrust through the tube just below the two open ends, as in fig. 10, in a tall glass jar or beaker filled with distilled water, so that the two open ends are above the surface of the water. The salts will diffuse readily (test for chlorides by nitrate of silver and nitric acid), but on applying any of the above tests no proteid will be found in the diffusate. (Peptones, however, are very diffusible.)

N.B.—The reactions d, e, f, and g are not obtained with peptones.

Preparation of Millon's Reagent.—Dissolve mercury in its own weight of strong nitric acid, specific gravity 1·4, and to the solution thus obtained add two volumes of water. Allow it to stand, and afterwards decant the clear fluid; or take one part of mercury, add two parts nitric acid, specific gravity 1·4, in the cold, and heat over a water-bath till complete solution occurs. Dilute with two volumes of water, and decant the clear fluid after twelve hours.

2. Presence of Nitrogen and Sulphur in Albumin.

(*a.*) Place some powdered dried albumin in a reduction tube, and into the mouth of the tube insert (1) a piece of red litmus paper, and (2) a lead acetate paper. On heating the tube the former becomes blue from the escape of ammonia, which can also be smelt, and the latter black from the formation of lead sulphide.

(*b.*) Heat some dry proteid with excess of soda-lime in a hard dry tube, when vapour of ammonia is evolved.

(*c.*) Place a few grains of the dry proteid, with a small piece of metallic sodium, in a dry hard tube, and heat slowly at first, and then strongly. After cooling, add carefully 3 cc. of water to the NaCy residue, filter, and to the filtrate add a few drops of ferric chloride and ferrous sulphate, and then add excess of hydrochloric acid. If nitrogen be present, there is a precipitate of Berlin blue, sometimes only seen after standing for a time.

3. Determination of Temperature of Coagulation.—"A glass beaker containing water is placed within a second larger beaker also containing water, the two being separated by a ring of cork. Into the water contained in the inner beaker there is immersed a test-tube, in which is fixed an accurately graduated thermometer, provided with a long narrow bulb. The solution of the proteid, of which the temperature of coagulation is to be determined, is placed in the test-tube, the quantity being just sufficient to cover the thermometer bulb. The whole apparatus is then gradually heated, and the experimenter notes the temperature at which the liquid first shows signs of opalescence" (*Gamgee*).

4. Circumstances Modifying the Coagulating Temperature.— Place 5 cc. of the solution of albumin in each of three test-tubes, colour them with litmus, and label them A, B, C. To A add a drop of very dilute acetic acid (0·1 per cent. acetic acid diluted five or six times); to B add a very dilute solution of caustic soda (0·1 per cent. of soda or potash similarly diluted); C is neutral for comparison. Place all three tubes in a beaker with water and heat them gradually, noting that coagulation occurs first in A, next in C, and last of all in B, the alkaline solution.

CLASSIFICATION AND PROPERTIES OF THE CHIEF PROTEIDS.

5. I.—Native Albumins are *soluble in water*, and are not precipitated by alkaline carbonates, sodic chloride, or very dilute acids. They are coagulated by heat at 65° to 73° C. When dried at 40° C. they yield a clear yellow, amber-coloured, friable mass "**soluble albumin**," which is soluble in water.

(1.) **Egg-Albumin.**—Prepare a solution as directed in Lesson I., 1. In addition, perform the following experiments :—

(*a.*) Evaporate some of the fluid to dryness at 40° C. over a water-bath to obtain "**soluble albumin.**" Study its characters, notably its solubility in water. This solution gives all the tests of egg-albumin. It is more convenient to purchase this substance.

(*b.*) Precipitate portions of the fluid with strong mineral acids, including sulphuric and hydrochloric acids.

(*c.*) Precipitate other portions by each of the following :— Mercuric chloride ; basic lead acetate ; tannic acid ; alcohol ; picric acid (see Lesson XVIII.)

(*d.*) Take 5 cc. of the fluid ; add twice its volume of 0·1 per cent. sulphuric acid, and then add ether. Shake briskly = coagulation after a time, at the line of junction of the fluids.

(*e.*) The solution is *not precipitated* on saturation with sodic chloride or magnesic sulphate (compare "Globulins ").

(2.) **Serum-Albumin.**— Dilute blood serum until it has the same specific gravity as the egg-albumin solution. Neutralise the solution with very dilute acid until a faint haziness is obtained.

Repeat all the tests for egg-albumin, but, in addition, note that the two solutions differ in the following respects :—

EGG-ALBUMIN.

SERUM-ALBUMIN.

(*a.*) Readily precipitated by hydrochloric acid, but the precipitate is not readily soluble in excess.

(*a.*) It is also precipitated by hydrochloric acid, but not so readily, while the precipitate is soluble in excess.

(*b.*) A non-alkaline solution is coagulated by ether.

(*b.*) It is not coagulated by ether.

(*c.*) The precipitate with nitric acid is soluble with difficulty in excess of the acid.

(*c.*) The corresponding precipitate is much more soluble in excess of acid.

(*d.*) The precipitate obtained by boiling is but slightly soluble in boiling nitric acid.

(*d.*) The corresponding precipitate is soluble in strong nitric acid.

[(*e.*) When injected under the skin, or introduced in large quantities into the stomach or rectum, it is given off by the urine.]

[(*e.*) When injected under the skin, it does not appear in the urine.]

6. II.—**Globulins** are insoluble in pure water, but are soluble in weak solutions of neutral salts—*e.g.*, sodic chloride—but insoluble in saturated ones. The solutions in these salts are coagulated by heat. They are soluble in dilute acids and alkalies, yielding acid- and alkali-albumins respectively. Most of them are precipitated from their saline solution by crystals of sodic chloride.

(1.) **Myosin,** see "Muscle."

(2.) **Serum-Globulin.**

(*a.*) Neutralise 5 cc. of blood-serum with a few drops of dilute sulphuric acid (0·1 per cent.), and then add 75 cc. of distilled water, and allow the precipitate to settle. Pour off the fluid and divide the precipitate into two portions, noting that it is insoluble in water, but soluble in excess of acid.

(*b.*) Boil a portion of the neutralised fluid = coagulation.

(*c.*) To 5 cc. of blood-serum in a test-tube, add an excess of *crystals* of magnesic sulphate, and shake briskly for some time. On standing, a white precipitate of serum-globulin falls.

Pour off the supernatant fluid, and observe that the precipitate is redissolved on the addition of water. Filter the supernatant fluid and test it for other proteids, which it still contains—viz., serum-albumin.

(*d.*) Take 5 cc. of blood-serum and pour a saturated solution of magnesic sulphate down the side of the glass to form a layer at the bottom of the tube. Where the two fluids meet, there is a copious white deposit of serum-globulin.

(*e.*) Treat another portion of serum to saturation with *crystals* of sodic chloride, and observe the same results.

(*f.*) Take another portion of serum, precipitate the serum-globulin with magnesic sulphate, and filter. To the filtrate add powdered sodic sulphate in excess, which gives a further precipitate. The filtrate still gives the reactions for proteids.

(**3.**) **Fibrinogen**, see " Blood."

7. III.—Derived Albumins are insoluble in pure water and in solutions of sodic chloride, but readily soluble in dilute hydrochloric acid and dilute alkalies. The solutions are not coagulated by heat.

(**1.**) **Alkali-albumin, or Alkali-albuminate—Preparation of Solution.**—Prepare a 5 per cent. solution of egg-albumin, as directed in Lesson I., 1.

(*a.*) To the egg-albumin add a few drops of solution of caustic soda or potash (0·1 per cent.), and heat it gently for a few minutes = **alkali-albumin.** Boil the fluid; it does not coagulate.

(*b.*) Test the reaction, it is alkaline.

(*c.*) Cool some of the alkali-albumin and colour it with litmus solution. Neutralise it carefully with very dilute acid = a precipitate on neutralisation, which is soluble in excess of the acid.

(*d.*) Repeat (*c.*); but, before doing so, add a few drops of sodic phosphate solution (10 per cent.), and note that the alkaline phosphates prevent the precipitation on neutralisation, until at least sufficient acid is added to convert the basic phosphate into acid phosphate. (See " Casein " under " Milk.")

(*e.*) Precipitate it by saturating its solution with crystals of common salt.

(*f.*) Lieberkühn's Jelly is really a strong solution of alkali-albumin. Place some undiluted white of egg in a test-tube, and add, drop by drop, a strong solution of caustic potash. The whole mass becomes stiff and glue-like, so that the tube can be inverted without the mass falling out.

(2.) Acid-Albumin or Syntonin.

Preparation.—(A.) To a 5 per cent. solution of egg-albumin add a few drops of dilute acid (*e.g.*, 0·1 per cent. sulphuric acid, or hydrochloric acid ·2 per cent.), and warm gently for several minutes = acid-albumin.

(B.) To finely-minced muscle, free from fat, add ten times its volume of dilute hydrochloric acid (4 cc. of acid in 1 litre of water), and allow it to stand for several hours, taking care to stir it frequently ; filter, the filtrate is a solution of syntonin.

(C.) Allow concentrated hydrochloric acid to act on fibrin, for a time, and filter.

Use the clear filtrate from A or B for testing.

(*a.*) The reaction is acid.

(*b.*) Boil the solution, it does not coagulate.

(*c.*) Neutralise another portion with very dilute potash or soda. A precipitate occurs, which is soluble in excess of the alkali. Employ litmus as on previous occasions.

(*d.*) Repeat (*c.*), but add sodic phosphate before neutralising; the syntonin is precipitated as before.

(*e.*) Add strong nitric acid = a precipitate which dissolves on heating, producing an intense yellow colour.

(*f.*) It also gives the biuret test, and that with Millon's reagent (Lesson I., 1).

8. IV.—Fibrin is insoluble in water and in weak solutions of common salt. When prepared from blood and washed, it is a white, fibrous, soft, and very elastic substance, which exhibits fibrillation under a high magnifying power (see " Blood ").

(*a.*) Place some well-washed fibrin in a test-tube, add some 0·1 per cent. hydrochloric acid, and observe that the fibrin swells up and becomes clear in the cold, but does not dissolve.

(*b.*) Place a test-tube as in (*a.*) on a water-bath at 60°C. for several hours; filter, and test the filtrate for acid-albumin by neutralisation with very dilute potash.

(*c.*) For the effect of a dilute acid and pepsin (see "Digestion ").

(*d.*) It decomposes hydric peroxide, and turns freshly-prepared tincture of guaiacum blue (see " Blood ").

(*e.*) Place a very dilute solution of cupric sulphate in a test-tube, add a flake of fibrin. The latter becomes greenish, while the fluid is decolourised. On adding caustic potash, the flake becomes violet. This is merely the biuret reaction common to proteids generally.

9. V.—Coagulated Proteids are insoluble in water, dilute acids, and alkalies, and are dissolved when digested at 35° to 40°C. in gastric juice (acid medium), or pancreatic juice (alkaline medium), forming peptones. They give Millon's reaction.

Preparation.—Boil white of egg hard, and chop up the white.

(*a.*) Test its insolubility in water, dilute acids, and alkalies.

(*b.*) It is partially soluble in acids and alkalies, when boiled for some time.

(*c.*) Bruise some of the solid boiled white of egg, diffuse it in water, and test it with Millon's reagent.

(*d.*) For the effect of the digestive juices see "Digestion."

10. VI.—Peptones are exceedingly soluble in water. Their solutions are not precipitated by sodic chloride, acids, or alkalies, nor are they coagulated by heat. They are precipitated by tannic acid, and with difficulty by excess of absolute alcohol.

Preparation (see "Digestion").—For applying the tests, dissolve a small quantity of Darby's Fluid Meat in water, and filter, or dissolve some pure peptone in water. The latter can be bought as a commercial product.

(*a.*) Boil a portion, it is not coagulated.

(*b.*) To another portion add strong nitric acid; and boil = a faint yellowish colour; allow it to cool, and add strong ammonia = orange colour (**Xanthoproteic Reaction**).

(*c.*) Acidify a third portion strongly with acetic acid, and add ferrocyanide of potassium = no precipitate.

(*d.*) Test separate portions with tannic acid, mercuric chloride, picric acid, and lead acetate. Each of these causes a precipitate. In the case of picric acid the precipitate disappears on heating, and partly reappears on cooling.

(*e.*) To another portion add a few drops of *very dilute* solution of cupric sulphate, and then caustic soda (or potash) =a *rose* colour; on adding more cupric sulphate, it changes to a violet (**Biuret Reaction**).

(*f.*) To another portion add a drop or two of Fehling's solution = a *rose* colour; on adding more Fehling's solution it changes to violet (**Biuret Reaction**).

(*g.*) Neutralise another portion = no precipitate.

(*h.*) To another portion add an excess of absolute alcohol = a precipitate of peptone, but not in a coagulated form.

(*i.*) Precipitate a portion with ferric acetate.

11. Diffusibility of Peptones.—Place a solution of peptones in a dialyser covered with an animal membrane, as directed in Lesson I., 1, (*h.*), and test the diffusate after some time for peptones.

THE ALBUMENOIDS.

12. I. Gelatin is obtained by the prolonged boiling of connective tissues, and from the hypothetical substance " **Collagen,**" of which fibrous tissue is said to consist.

Preparation of a Solution.—Use commercial gelatin. Make a watery solution by allowing it to swell up in water, and then dissolving it with the aid of heat.

(*a.*) It is insoluble, but swells up, in cold water.

(*b.*) After a time heat the gelatin swollen up in water; it dissolves. Allow it to cool; it gelatinises.

(*c.*) Precipitate separate portions by each of the following: Mercuric chloride, tannin, alcohol, and platinic chloride.

(*d.*) It is not precipitated by acids (acetic or hydrochloric), or alkalies, or lead acetate.

(*e.*) It is not precipitated by acetic acid and potassic ferrocyanide (unlike albumin).

(*f.*) It is not coagulated by heat (unlike albumin).

(*g.*) It gives the xanthoproteic and biuret tests, and that with Millon's reagent. It is precipitated by picric acid; the precipitate is dissolved on heating, and reappears on cooling.

13. II. Chondrin is derived by prolonged boiling from the matrix of cartilage, which is supposed to consist of the hypothetical substance "Chondrogen." It seems to be really a mixture of mucin and glutin.

14. III. Mucin, see " Bile."

LESSON II.

THE CARBOHYDRATES, FATS, BONE.

1. I. Starch $(C_6H_{10}O_5)_n$ **Preparation.** — Wash a potato thoroughly, and grate it on a grater into water in a tall cylindrical glass. Allow the suspended particles to subside, and after a time note the deposit ; the lowest stratum consists of a white powder or starch, and above it lie coarser fragments of cellulose and other matters. Decant off the supernatant fluid.

(*a.*) **Microscopical Examination.**—Examine the white deposit of starch, noting that each starch granule shows an eccentric hilum with concentric markings. Add a very dilute solution of iodine. Each granule becomes blue, while the concentric markings become more distinct.

(*b.*) At this stage it is advantageous to compare the microscopic characters of other varieties of starch—*e.g.*, rice, arrowroot, &c. (Fig. 1).

Fig. 1.—*e*, Tahiti arrowroot ; *d*, Potato starch.

(*c.*) **Polariscope.**—Examine starch granules with a polarisation-microscope. With crossed Nicol's, when the field is dark, each granule shows a dark cross on a white refractive ground.

(*d.*) Squeeze some dry starch powder between the thumb and forefinger, and note the peculiar crepitation sound and feeling.

2. Prepare a Solution.—Place 1 grm. of starch in a mortar, rub it up with a little cold water, and then add 50 cc. of boiling water, and rub up the whole together until the starch is apparently dissolved and a somewhat opalescent fluid is obtained. Allow the solution to cool. [In reality, the starch is only imperfectly dissolved by hot water.]

(*a.*) Add powdered dry starch to cold water. It is insoluble. Filter, and test the filtrate with iodine. It gives no blue colour.

(*b.*) The above 'method shows that it is imperfectly dissolved in warm water. If more starch be used, a thick " starch-paste," which sets on cooling, is obtained.

(*c.*) To a portion of the above fluid add a solution of iodine = a blue colour, which disappears on heating and reappears on cooling—provided it has not been boiled too long. Place the test-tube in cold water to cool it.

(*d.*) Render some of the starch solution alkaline by adding caustic soda solution. Add iodine solution. No blue colour is obtained.

(*e.*) Acidify (*d.*) with dilute sulphuric acid, then add iodine = blue colour is obtained.

(*f.*) To another portion of the solution, add a few drops of dilute cupric sulphate and caustic soda (or Fehling's solution), and boil = no reaction (compare " Grape Sugar ").

(*g.*) Add tannic acid = yellowish precipitate, which dissolves on heating.

3. Starch is a Colloid.—Place some strong starch solution in a dialyser or parchment-tube, and the latter in distilled water. Allow it to stand for some time, and test the water for starch ; none will be found.

4. II. Dextrin ($C_6H_{10}O_5$).

Prepare a Solution.—Dissolve some dextrin in boiling water, and observe that the solution is not opalescent.

(*a.*) This proves its solubility in water.

(*b.*) To a portion of the solution, add a solution of iodine = reddish-brown colour, which disappears on heating and returns on cooling. [The student ought to use two test-tubes, placing the dextrin solution in one, and an equal volume of water in the other. Add to both an equal volume of solution of iodine, and thus compare the difference in colour.]

(*c.*) Precipitate some from its solution by adding alcohol.

(*d.*) Render some of the dextrin solution alkaline by adding caustic soda solution. No red-brown colour is obtained with iodine. Acidify and the reddish-brown colour appears.

5. III.—Glycogen or Animal Starch, $C_6H_{10}O_5$.

Prepare a Solution (see " Liver ").

(*a.*) Take a portion of the solution ; note its *opalescence ;* add solution of iodine (made by adding iodine to water in which a crystal of potassic iodide is dissolved) = red-brown or port-wine-red colour. As in the dextrin test, use two test-tubes, one with water and the other with glycogen, to compare the difference in colour. The colour disappears on heating, and reappears on cooling.

6. IV.—Glucose, Dextrose, or Grape-Sugar, $C_6H_{12}O_6$.

In commerce it occurs in warty uncrystallised masses, of a yellowish or yellowish-brown colour. It is readily soluble in water. Prepare a solution by dissolving a small quantity in water.

(*a.*) To the solution add iodine = no reaction.

(*b.*) To the solution add a trace of a dilute solution of cupric sulphate, and afterwards add caustic soda (or potash) until the precipitate first-formed is redissolved, and a clear blue fluid is obtained. Boil gradually ; if grape-sugar be present, the blue colour disappears, and a yellow precipi-tate of hydrated cuprous oxide is obtained. It is well to boil the surface of the fluid, and when the yellow precipitate occurs, it contrasts sharply with the deep blue-coloured stratum below. If no sugar be present, only a black colour may be obtained (**Trommer's Test**).

(*c.*) To the solution add Fehling's solution; boil = a yellow or yellowish-red precipitate of hydrated cuprous oxide.

[For the precautions to be observed in using Fehling's solution, and for other tests for glucose, see " Urine."]

7. **V.—Maltose, $C_{12}H_{22}O_{11}$.**

(*a.*) Take 1 grm. of ground malt, mix it with ten times the quantity of water, place the mixture in a beaker, and keep it at 60°C. for half an hour. Then boil and filter ; the filtrate contains maltose and dextrin.

(*b.*) Test for a reducing sugar with Fehling's solution or other suitable test. (See also " Salivary digestion.")

8. **VI.—Lactose, $C_{12}H_{22}O_{11}$ + H_2O** (see " Milk ").

9. **VII.—Cane Sugar, $C_{12}H_{22}O_{11}$.**

(*a.*) Observe its crystalline form and sweet taste.

(*b.*) Its solutions do not reduce Fehling's solutions (many of the commercial sugars, however, contain sufficient grape-sugar to do this).

(*c.*) Place some cane-sugar in a beaker, pour on it strong sulphuric acid, and add a few drops of water ; soon the whole mass is charred.

(*d.*) **Inversion of Cane-Sugar.**—Boil a strong solution of cane-sugar in a flask with one-tenth of its volume of strong hydrochloric acid. After prolonged boiling the cane-sugar is "inverted," and the solution contains a mixture of dextrose and lævulose. Test its reducing-power with Fehling's solution.

10. **Conversion of Starch into a Reducing Sugar.**—Place 50 cc. of starch solution in a flask on wire gauze over a Bunsen burner, add one drop of strong sulphuric acid, and boil for five to ten

minutes, observing the spluttering that occurs, the liquid meantime becoming limpid.

(*a.*) Test a portion of the liquid for glucose, taking care that sufficient alkali is added to neutralise the surplus acid.

(*b.*) Test it with iodine = a blue colour, showing that some soluble starch (amidulin) still remains unconverted into a reducing sugar.

11. Circumpolarisation.—Certain substances when dissolved possess the power of rotating the plane of polarised light—*e.g.*, the proteids, sugars, &c. The extent of the rotation depends on the amount of the active substance in solution. The direction of rotation—*i.e.*, to the right or the left—is constant for each active substance. Of course, light of the same wave-length must be used. The light obtained from the volatilisation of common salt is used.

The term "specific rotatory power," or "specific rotation" of a substance, is used to indicate the amount of rotation expressed in degrees of the plane of polarised light, which is produced by 1 grm. of the substance dissolved in 1 cc. of liquid, when examined in a layer 1 decimetre thick.

Those substances which cause specific rotation are spoken of as "*optically active ;*" those which do not, as "*inactive.*"

If a = the observed rotation ;
p = the weight in grammes of the active substances contained in 1 cc. of liquid ;
l = the length of the tube in decimetres ;
$(a)_D$ = the specific rotation for light corresponding to the light of a sodium flame ;
then

$$(a)_D = \pm \frac{a}{p\,l}$$

The sign + or – indicates that the substance is dextro- or lævo-rotatory. Various instruments are employed. Use

Laurent's Polarimeter.—This instrument must be used in a dark room.

12. Determination of the Specific Rotatory Power of Dextrose.

(*a.*) Fill one of the decimetre tubes with distilled water, taking care that no air-bubbles get in. Slip on the glass disc horizontally, and screw the brass cap on the tube. Place the tube in the instrument, so that it lies in the course of the rays of polarised light.

(*b.*) Place some common salt (or fused common salt, and sodic carbonate) in the platinum spoon, and light the Bunsen's lamp, so that the soda is volatilised. If a platinum spoon is not available, tie several platinum wires together, dip them into slightly moistened common salt, and fix them in a suitable holder, so that the salt is volatilised in the outer part of the flame. In the newer form of the instrument supplied by Laurent, there are two Bunsen burners, placed the one behind the other, which give very much more light. Every part of the apparatus must be scrupulously clean.

(*c.*) Bring the zero of the vernier to coincide with that of the scale. On looking through the eye-piece, and focussing the vertical line dividing the field vertically into two halves, the two halves of the field should have the same intensity when the scale reads zero. If this is not the case, then adjust the prisms until it is so, by means of the milled head placed for that purpose behind the index dial and above the telescope tube. It is well to work with the field not too brightly illuminated.

(*d.*) Remove the water tube, and substitute for it a similar tube containing the solution of the substance to be examined —in this case a *perfectly clear solution* of pure dextrose. Place the tube in position, and proceed as before. The two halves of the field are now of unequal intensity. Rotate the eye-piece until equality is obtained.

(*e.*) Repeat the process several times, and take the mean of the readings. The difference between this reading and the first at (*c.*), when the tube was filled with distilled water —*i.e.*, zero—is the rotation due to the dextrose = (*a.*)

(*f.*) Place 10 cc. of the solution of dextrose in a weighed capsule, evaporate to dryness over a water-bath, let the capsule cool in a desiccator, and weigh again. The increase in weight gives the amount of dextrose in 10 cc.; so that the amount in 1 cc. is got at once = p.

(*g*.) Calculate the specific rotatory power by the above formula. It is about + 53°.

For practice, begin with a solution of dextrose containing 11 grms. per 100 cc. of water. Make several readings of the amount of rotation, and take the mean.

Example.—In this case, the mean of the readings was 11·6°

$$a_D = \frac{11 \cdot 6°}{\cdot 11 \times 2} = 53°$$

Repeat the process with a 4 and 2 per cent. solution. It is necessary to be able to read to 2 minutes, but considerable practice is required to enable one to detect when the two halves of the field have exactly the same intensity.

Test the rotatory power of corresponding solutions of cane-sugar, and any other sugar you please.

Test also the rotatory power of a proteid solution.

The following indicate the S.R. for yellow light :—

Proteids.—Egg-albumin - 35·5°; serum-albumin - 56°; syntonin - 72°; alkali-albumin prepared from serum-albumin - 86°, when prepared from egg-albumin − 47°.

Carbohydrates. — Glucose + 56°; maltose + 150°; lactose + 52·5°.

NEUTRAL FATS.

13. Reactions.

(*a*.) Use almond oil or lard, and observe that fat is soluble in ether, chloroform, and hot alcohol.

(*b*). To almond oil add caustic soda, and boil = **saponification.**

(*c*.) Shake oil containing a fatty acid—*e.g.*, De Jongh's cod-liver oil, with a few drops of a dilute solution of sodic carbonate. The whole mass becomes white = **emulsion.** Examine it microscopically, and compare it with milk, which is a typical emulsion.

(*d.*) Shake up olive oil with a solution of albumin in a test-tube = an emulsion. Examine it microscopically.

(*e.*) Heat in a porcelain capsule for an hour or more some lard mixed with plumbic oxide and a little water. The fat is split up, yielding glycerin and a lead soap.

(*f.*) Heat some lard and caustic soda solution in a capsule to form a soap; decompose the latter by heating it with dilute sulphuric acid, and observe the liberated fatty acids floating on the top.

BONE.

14. A.—Organic Basis of Bone.

(*a.*) To Decalcify Bone.—Place a small thin dry bone in dilute hydrochloric acid (1 : 8) for a few days. Its mineral matter will be gradually dissolved out, when the bone, although retaining its original form, loses its rigidity, and becomes pliable, elastic, and so soft as to be capable of being cut with a knife. What remains is the organic matrix or ossein. Keep the solution obtained.

(*b.*) Wash the decalcified bone thoroughly with water, in which it is insoluble. Boil it for a long time, and from it gelatin will be obtained. Test the solution for gelatin (Lesson I., 12).

B.—Mineral Matter in Bone.

(*a.*) Examine a piece of bone which has been incinerated in a clear fire. At first the bone becomes black from the carbon of its organic matter, but ultimately it becomes white. What remains is calcined bone, having the form of the original bone, but now it is quite brittle. Powder some of the white bone-ash.

(*b.*) Dissolve a little of the powdered bone-ash in hydrochloric acid, observing that bubbles of gas (CO_2) are given off, indicating the presence of a carbonate; dilute the solution, add excess of ammonia = a white precipitate of phosphate of lime and phosphate of magnesia.

(*c.*) Filter, and to the filtrate add ammonium oxalate = a white precipitate of oxalate of lime, showing that there is lime present, but not as a phosphate.

(*d.*) To the solution of mineral matters 14, A. (*a.*) add acetate of soda until there is free acetic acid present, recognised by the smell; then add ammonium oxalate = a copious white precipitate of lime salts.

Exercises on the Foregoing.—The solution may contain one or more proteids or carbohydrates.

A. Proteids.

(1.) Note the colour, odour, and transparency of the solution.

(2.) Test its reaction. Neutralise by dilute sodic carbonate or hydrochloric acid, if necessary. If the precipitate gives the xanthoproteic reaction, it is acid or alkali-albumin. If not, it is earthy phosphates.

(3.) Do the xanthoproteic reaction, which shows the presence of a proteid.

(4.) Boil. If there is coagulation it is either egg-albumin, serum-albumin, or globulin.

(*a.*) Test if the solution is precipitated by crystals of magnesic sulphate = globulin. Filter.

(*b.*) Test the filtrate of (*a.*) by acidulation and heat for albumin. Confirm by other tests.

(*c.*) Test the filtrate of (*b.*) for peptones.

(5.) Test for gelatin.

B. Carbohydrates.

(1.) Remove any proteids present, except peptones, by acidification and boiling, and use this solution for testing.

(2.) Add iodine after acidulation if necessary, a blue colour = starch; a port-wine colour = dextrin or glycogen. Confirm by other tests.

(3.) If no peptones are present, test for sugar.

(4.) If peptones are present, evaporate to dryness, dissolve the residue in 98 per cent. alcohol, filter. Evaporate the alcohol and redissolve the residue in water, and test for sugar.

LESSON III.

THE BLOOD—COAGULATION, ITS PROTEIDS.

1. Reaction.—Prick your finger with a needle and place a drop of the freshly shed blood on a strip of dry, smooth, *glazed*, neutral litmus paper. Allow it to remain for a short time ; then wash it off with a stream of distilled water from a wash-bottle, when a blue spot upon a red or violet ground will be seen, indicating its *alkaline* reaction.

2. Blood is Opaque.

(*a*). Place a thin layer of defibrinated blood on a microscopic slide, and try to read some printed matter through it. This will be found impracticable.

3. To make Blood Transparent or Laky.—Place 10 cc. of the defibrinated blood provided for you in three test-tubes, labelling them A, B, and C. Keep A for comparison.

(*a.*) To B add 5 volumes of water, and warm slightly, noting the change of colour by reflected and transmitted light. When looked at by reflected light, it is much darker in colour, in fact, it looks almost black, but by transmitted light it is transparent. Test this by looking as in 2 (*a.*) at printed matter.

(*b.*) To C add a solution of taurocholate of soda. Test its transparency as above. In 2, the hæmoglobin is still within the blood corpuscles. In the others—3 (*a.*), (*b.*)—it is dissolved out, and in solution.

4. Action of Saline Solution.

(*a.*) Take 2 cc. of defibrinated blood in a test-tube, label it D, add 5 volumes of a 10 per cent. solution of sodic chloride. Observe the change of colour. It becomes a very bright florid red, more brick-red than the original blood itself. Compare its colour with that in A, B, and C. It is opaque.

5. Hæmoglobin does not Dialyse.

(*a.*) Place a watery solution of defibrinated blood in a dialyser, and suspend it in a large vessel of distilled water. Carefully test the dialyser beforehand to see that there are no holes in it. If there are any fine pores, close them with a little white of egg, and coagulate it with a hot iron.

(*b.*) After several hours observe that no hæmoglobin has passed into the water.

(*c.*) Test the diffusate for chlorides.

6. Phenomena of Coagulation.—Place a small porcelain capsule on the table ; decapitate a rat, and allow the blood to flow into the capsule. Within a few minutes the blood **congeals,** and when the vessel is tilted the blood no longer flows as a fluid, but as a **solid.** It then **coagulates** completely. Allow it to stand, and after an hour or so, pale-yellow coloured drops of a fluid—the **serum**—are seen on the surface, being squeezed out of the red mass, the latter being the **clot.**

7. Frog's Blood—Coagulation of the Plasma.—Place 5 cc. of normal saline (0·75 per cent. salt solution) in a test-tube surrounded with ice. Expose the heart of a pithed frog, and cut into the ventricle, allowing the blood as it escapes to flow into the normal saline. Mix the two, and the corpuscles (owing to their greater specific gravity) after a time subside. After they have subsided remove the supernatant fluid—the plasma mixed with normal saline—by means of a pipette. Place it in a watch-glass, and observe that it coagulates.

8. Mammalian Blood.

(**A.**) Study coagulated blood obtained from the slaughter-house. Run the blood of a sheep or ox into a tall cylindrical vessel, and allow it to coagulate. Set it aside for two days, and then observe the **serum** and the **clot.** Pour off the pale, straw-coloured serum, and note the red clot, which has the shape of the vessel, although it is smaller than the latter.

(**B.**) If the blood of a **horse** can be obtained, study it, noting that the upper layer of the clot is paler in colour; this is the **buffy coat.**

9. Circumstances influencing Coagulation.

Effect of Cold.—Place a small platinum basin—a brass or glass thimble will do quite well—on a freezing mixture of ice and salt, decapitate a frog or rat, and allow the blood to flow directly into the cooled vessel. At once it becomes **solid or congeals,** but it is not coagulated. As soon as the blood becomes solid, remove the thimble and thaw the blood by placing it on the warm palm of the hand, when the blood becomes fluid, so that it can be poured into a watch-glass; if the vessel be once more placed on the freezing mixture, the blood again congeals and solidifies, and on its being removed becomes fluid. Observe at the same time that the colour of the blood changes, becoming darker and transparent. This is the *laky* condition due to the discharge of the hæmoglobin from the corpuscles. Place the vessel with the fluid blood on the table, and it clots or forms a firm jelly.

10. Influence of Neutral Salts on Coagulation.—Take to the slaughter-house a vessel capable of holding 500 cc., but previously place in it 170 cc. of a saturated solution of sodic sulphate or magnesic sulphate. Allow enough blood from an animal to run into the saline solution to fill the vessel, and mix them thoroughly. The blood does not clot but remains fluid. Place the vessel aside on ice, and note that the corpuscles subside, leaving a clear yellowish layer on the surface—the plasma mixed with the saline solution, and known as **salted plasma.**

(*a.*) Pipette off the salted plasma—use 2 cc.—add to it 3 to 5 volumes of water, and observe that it clots after a time. The clotting is hastened by the action of gentle heat.

In laboratories where a centrifugal apparatus is in use, the corpuscles can be rapidly separated from the plasma, and enough of the latter obtained for the purposes of a large class of students.

(*b.*) Place 15 cc. of the salted plasma—separated by means of the centrifugal apparatus—in a tall, narrow, cylindrical, stoppered glass. Add powdered sodic chloride, and shake the whole vigorously, when a white flocculent precipitate is thrown down. Allow the precipitate to subside. Decant off the supernatant fluid and the salt solution. Filter through a filter, moistened with a *saturated* solution of sodic chloride, and wash the precipitate on the filter with saturated solution of sodic chloride. This is the plasmine of Denis. Scrape the washed precipitate off the filter by means of a knife.

Dissolve it in a small quantity of distilled water, and filter quickly. The filtrate if set aside will clot after a time. It is better to do the several operations rapidly to ensure success, but I have frequently found coagulation occur when the plasmine was not dissolved in water until many hours after it was deposited.

11. Defibrinated Blood.—In the slaughter-house allow blood to run from an animal into a vessel, and with a bundle of twigs beat or whip the blood steadily for some time. Fine white fibres of fibrin collect on the twigs, while the blood remains fluid. This is **defibrinated blood**, and although set aside for any length of time, it does not coagulate spontaneously.

(*a.*) With a few thin twigs, or the barbed end of a quill, beat some freshly-shed blood, and observe the fibrin sticking to the twigs. Wash it.

12. I.—Fibrin.—Take the twigs coated with fibrin of the previous experiment. Wash away the colouring-matter with a stream of water until the fibrin becomes quite white.

(*a.*) Study its **physical properties** : it is a white, fibrous, highly-elastic substance. Stretch some fibres to observe their extensibility; on freeing them, they regain their shape, showing their elasticity.

(*b.*) Place a few fibrils in absolute alcohol to rob them of water, when they become brittle, and lose their elasticity.

(*c.*) Place a flake in a test-tube with some 0·2 per cent. hydrochloric acid in the cold. It swells up and becomes clear and transparent, but does not dissolve.

(*d.*) Repeat (*c.*), but place the test-tube in a water-bath at 60° C., the fibrin is dissolved forming acid-albumin. Test for the latter (Lesson I., 7, III., 2).

(*e.*) Place a few fibrils in a watch-glass, and pour over them some hydric peroxide; bubbles of oxygen are given off. Immerse a flake in freshly-prepared tincture of guaiacum (5 per cent. solution of the pure resin in alcohol), and then in hydric peroxide, when a blue colour is developed. If the fibrin contain much water, it is preferable to place it first of all for a short time in rectified spirit to remove the water. [Other substances give a blue colour under similar conditions.]

(*f.*) Suspend some fibrils of fibrin in water in a test-tube, and observe that they give (1) the Xanthoproteic reaction, and that with Millon's reagent (Lesson I., 1).

(*g.*) Prick a finger with a needle; collect a drop of blood on a microscopic slide, cover, and examine under a microscope (× 350). After a time observe the formation of threads of fibrin between the rouleaux of coloured blood-corpuscles.

13. II.—Blood-Serum.—By means of a pipette remove the serum from the coagulated blood (Lesson III., 8). If a centrifugal apparatus is available, any suspended blood-corpuscles can easily be separated by it. Observe its straw-yellow colour. Test its reaction; it is *alkaline.*

Study its **proteids**. Test first for the **general reactions** common to all proteids.

(*a.*) Dilute 1 volume of blood-serum with 50 volumes of water, and use this for testing.

(*b.*) Test separate portions by neutralisation and heat; nitric acid and the subsequent addition of ammonia; acetic acid and ferrocyanide of potassium; Millon's reagent, and the biuret reaction (Lesson I., 1). The solution gives all these reactions.

Study its **individual proteids.**

(**A.**) **Preparation of Serum-globulin (Para-globulin** or Fibrinoplastin).

(*a.*) Take 10 cc. of blood-serum ; add 200 cc. of ice-cold water, and pass a stream of carbon dioxide through it for some time = a white precipitate of serum-globulin. This method does not precipitate it entirely (**Schmidt's method**). No precipitate is obtained unless the serum be diluted.

(*b.*) Dilute 10 cc. of blood-serum with 150 cc. of water; add 5 drops of a 20 per cent. dilution of acetic acid = a white precipitate of serum-globulin, or as it was called, " serum-casein " (**Panum's method**). All the serum-globulin is not precipitated.

(*c.*) To 5 cc. of fresh blood-serum in a test-tube, add crystals of magnesium sulphate in large excess, and shake briskly for some time. The excess of crystals falls to the bottom, and on their surface is precipitated a dense white flocculent mass of serum-globulin (**Hammarsten's method**). Allow the excess of the salt and the precipitate to settle. Decant the bulk of the supernatant fluid, and filter the remainder. Wash the precipitate on the filter with a saturated solution of magnesic sulphate ; add a little distilled water to the precipitate. It is dissolved—*i.e.*, it is a globulin, and is insoluble in excess of a neutral salt, but is dissolved by a weak solution of the same. The solution does not coagulate spontaneously.

(*d.*) The solution obtained in (*c.*) gives all the reactions for proteids with the special reactions of a globulin.

(*e.*) Allow a few drops of blood-serum to fall into a large quantity of water, and observe the milky precipitate of a globulin = serum-globulin. This is best observed by placing a dead black surface behind the vessel of water. We can then trace the " milky way " of the falling drops of serum as they mix with the water.

(**B.**) **Serum-Albumin.**—From (**A.**), (*a.*), (*b.*), (*c.*), filter off the precipitate, and test the filtrate for the usual proteid reactions, so that the filtrate still contains a proteid which is serum-albumin (Lesson I., 5, 2).

14. Precipitation by other Salts.

(*a.*) Precipitate the serum-globulin of blood-serum with magnesic sulphate. Filter, and to the filtrate add sodic sulphate, when serum-albumin is precipitated. Sodic sulphate, however, gives no precipitate with pure serum.

(b). Precipitate blood-serum with potassic phosphate. All the proteids are thrown down after prolonged shaking.

(c.) Precipitate blood-serum with magnesic sulphate and sodic sulphate, or the double salt sodio-magnesic sulphate. All the proteids are thrown down.

15. Preparation of Fibrinogen.

(a.) Dilute 10 cc. of hydrocele fluid with 150 to 200 cc. of water, and pass through it for a considerable time a stream of carbon dioxide, when there is precipitated a small quantity of a somewhat slimy white body, fibrinogen.

(b.) Take 10 cc. of hydrocele fluid and add powdered crystals of common salt to saturation, as for the preparation of paraglobulin (Lesson III., 13, A.)

16. Coagulation Experiments.

(a.) Andrew Buchanan's Experiment.—Mix 5 cc. fresh blood-serum (preferably from horse's blood) with 5 cc. hydrocele fluid, and keep the mixture at 35°C. for some hours, when coagulation occurs, a clear pellucid clot of fibrin being obtained.

(b.) To 5 cc. of hydrocele fluid, add some solution of paraglobulin (prepared as in Lesson III., 13, A); coagulation will result after a time.

(c.) Modify (a.) in the following manner :—To 2 cc. of fresh blood-serum, add 2 cc. of a solution of fibrinogen (prepared as in Lesson IV., 15, b) = coagulation.

(d.) To 2 cc. of salted plasma, prepared as in Lesson III., 10 (which is known to clot slowly on the addition of water), add 10 volumes—i.e., 20 cc. of a watery solution of fibrin-ferment, prepared by the demonstrator = coagulation.

17. The Salts present in blood are to be tested for in the usual way.

18. Preparation of Fibrin-ferment.—It must be kept in stock. Precipitate blood-serum with a large excess of alcohol, collect the copious precipitate; cover it with absolute alcohol, and allow it to stand at least a month—the longer the better. Dry the precipitate at 35°C., and afterwards over sulphuric acid. Keep it as a dry powder in a well-stoppered bottle. When a solution is required, extract some of the dry powder with 100 volumes of water; filter. The filtrate contains the ferment.

LESSON IV.

THE COLOURED BLOOD-CORPUSCLES.

SPECTRA OF HÆMOGLOBIN AND ITS COMPOUNDS.

Enumeration of the Corpuscles.—Several forms of instruments are in use, *e.g.*, those of Malassez, Zeiss, and Gowers.

1. The Hæmocytometer of Gowers' consist of

(*a.*) A small pipette, which, when filled to the mark on its stem, holds 995 c.mm. (Fig. 2, A).

(*b.*) A capillary tube to hold 5 c.mm. (B).

(*c.*) A small glass jar in which the blood is diluted (D).

(*d.*) A glass stirring rod (E).

(*e*). Fixed to a brass plate a cell $\frac{1}{5}$ of a millimetre deep, and with its floor divided into squares $\frac{1}{10}$ mm., in which the blood-corpuscles are counted.

(*f.*) The diluting solution consists of a solution of sodic sulphate in distilled water—specific gravity, 1025.

This instrument can be used with any microscope.

2. Mode of using the Instrument.

(*a.*) By means of the pipette (A) place 995 c.mm. of the diluting solution in the mixing jar (D).

(*b.*) Puncture a finger near the root of the nail with the
lancet projecting from (F), and with the pipette (B) suck
up 5 c.mm. of the blood, and blow it into the diluting solution, and mix the two with the stirrer (E).

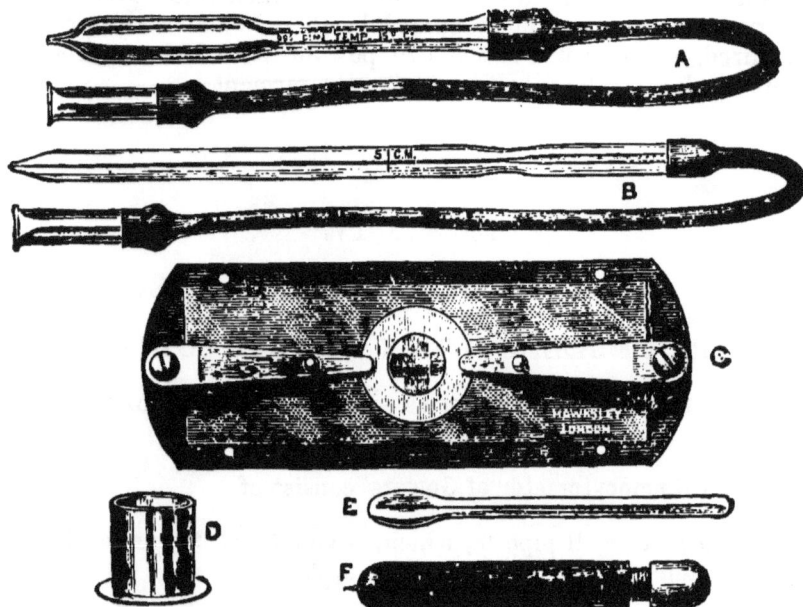

Fig. 2.—Gowers' Hæmocytometer.—A, Pipette for measuring the diluting
solution; B, for measuring the blood; C, cell with divisions on the
floor, mounted on a slide, to which springs are fixed to secure the
cover glass; D, vessel in which the solution is made; E, spud for
mixing the blood and solution; F, guarded spear-pointed needle.

(*c.*) Place a drop of the mixture on the centre of the glass
cell (C), see that it exactly fills the cell, and cover it gently
with the cover-glass, securing the latter with the two springs.
Place the cell with its plate on the stage of a microscope,
and focus for the squares ruled on its base.

(*d.*) When the corpuscles have subsided, count the
number in 10 squares, and this, when multiplied by 10,000,
gives the number in a cubic millimetre of blood

(*e.*) Wash the instrument, and in cleaning the cell do
this with a stream of distilled water from a wash-bottle.

Take care not to brush the cell with anything rougher than a camel's-hair pencil, to avoid injuring the lines.

Each square has an area of $\frac{1}{100}$ mm., so that 10 squares have an area of $\frac{1}{10}$ mm. As the cell is $\frac{1}{5}$ mm. deep, the volume of blood in 10 squares is $\frac{1}{10} \times \frac{1}{5} = \frac{1}{50}$ c.mm. On counting the number of corpuscles in 10 squares, and multiplying by 50, this will give the number in 1 c.mm. of the *diluted* blood. On multiplying this by $\frac{1000}{5}$, we get the number in 1 c.mm. before dilution. Thus we arrive at the reason why we multiply the number in 10 squares by 10,000 to get the number of corpuscles in 1 c.mm. of blood.

HÆMOGLOBIN AND ITS DERIVATIVES.

3. Preparation of Hæmoglobin Crystals.

(*a*.) **Rat's Blood.**—Place a drop of the defibrinated rat's blood provided for you on a slide, add three or four drops of water, mix, and cover with a cover-glass. Examine the slide with a high power of the microscope; after a few minutes, especially at the edges of the preparation, small crystals will begin to form, and gradually grow larger. The crystals are those of oxy-hæmoglobin, and have the form of thin rhombic plates, disposed singly or in groups.

(*b*.) **Dog's Blood.**—To 15 cc. of the defibrinated dog's blood provided for you, add, drop by drop, 1 cc. or so of ether, shaking the tube after each addition of ether. By this means the blood is rendered *laky*, a condition which is recognised by inclining the tube, and observing that the film of blood left on it, on bringing the tube to the vertical again, is transparent. Add no more ether, but place the tube in a freezing-mixture of ice and salt; as the temperature falls, crystals of hæmoglobin separate. If the crystals do not separate at once, keep the tube in the freezing-mixture for one or two days. Examine some of the crystals under the microscope.

4. Ozone Test for Hæmoglobin.—Mix some freshly-prepared alcoholic solution of guaiacum with ozonic ether; the mixture becomes turbid, and on adding even a dilute solution of hæmo-

globin, a blue colour results. Or the reaction may be done on filter paper.

5. Spectroscopic Examination of Blood.—Use a small Browning's straight-vision spectroscope (Fig. 3).

Fig. 3.—Browning's Straight Vision Spectroscope.

Preliminary.—Observe the solar spectrum by placing the spectroscope before the eye, and directing it to the bright daylight. Note the spectrum from the red to the violet end, with the intermediate colours, and particularly the dark Frauenhofer's lines, known as D, E, b, and F, their position and relation to the colours. Make a diagram of the colours, and the dark lines, indicating the latter by their appropriate letters.

(*a.*) Fix the spectroscope in a suitable holder, and direct it to a gas-flame, the edge of the flame being towards the slit in the spectroscope, noting that the spectrum shows no dark lines.

(*b.*) Fuse a piece of platinum wire in a glass tube, and make a loop at the free end of the platinum wire. Dip the platinum wire in water and then into common salt, and burn the salt in the gas-flame, having previously directed the spectroscope towards the gas-flame, and so arranged the latter that it is seen edge-on. Note the position of the bright yellow sodium line D.

6. I. Spectrum of Oxy-hæmoglobin.

(*a.*) Begin with a strong solution, and gradually dilute it. Place some defibrinated blood in a test-tube, and observe its opacity and bright scarlet colour.

(*b.*) Adjust the spectroscope as follows :—Light a fan-tailed gas-burner, fix the spectroscope in a suitable holder, and between the light and the slit of the spectroscope place a test-tube containing the blood or its solution. Focus the *long image* of the gas-flame on the slit of the spectroscope. The focal point can be readily ascertained by holding a sheet of white paper behind the test-tube.

(*c.*) Add 10 to 15 volumes of water, and note that only the red part of the spectrum is visible. Make a sketch of what you see, noting the dilution.

(*d.*) Add more water until the green appears, and observe that a single dark **absorption band** appears between the red and green (Fig. 4, 1). Continue to dilute until this single broad band is resolved into two by the transmission of yellow-green light. Burn a bead of sodic chloride in the gas-flame, to note distinctly the position of the D line, and observe that of the **two absorption bands** the one nearest D, conveniently designated by the letter α, is more sharply defined and narrower; while the other, conveniently designated by the letter β, nearer the violet end, is broader and fainter. At the violet end the spectrum is shortened by absorption (Fig. 4, 2).

(*e.*) Continue to dilute the solution, and note that the band near the violet end is the first to disappear.

Sketch the appearances seen with each dilution, and indicate opposite each the degree of the latter.

(*f.*) A very instructive method is to make a pretty strong solution of blood, showing only one undivided band. Place a little of this in a test-tube, and pour in water, so that the water mixes but slightly with the upper strata of the blood. Examine the solution spectroscopically, moving the tube so as to examine first the deeper strata of fluid until the surface is reached. At first a single band is seen; but as the solution is more dilute above, the two bands begin to appear, and as the solution gets weaker above, the β-band disappears, until, finally, with a very weak solution, such as is obtained in the upper strata only, the α-band alone is visible.

7. Hæmatinometer.—For accurate observation, instead of a test-tube the blood is introduced into a vessel with parallel sides, the glass plates being exactly 1 cm. apart (Fig. 6, D). Study this instrument (*Hoppe-Seyler*).

8. Hæmatoscope.—By means of this instrument the depth of the stratum of fluid to be investigated can be varied, and the variation of the spectrum, with the strength of the solution, or the thickness of the stratum through which the light passes, at once determined (*Hermann*). Study this instrument.

Red. Orange.   Yellow.   Green.   Blue.

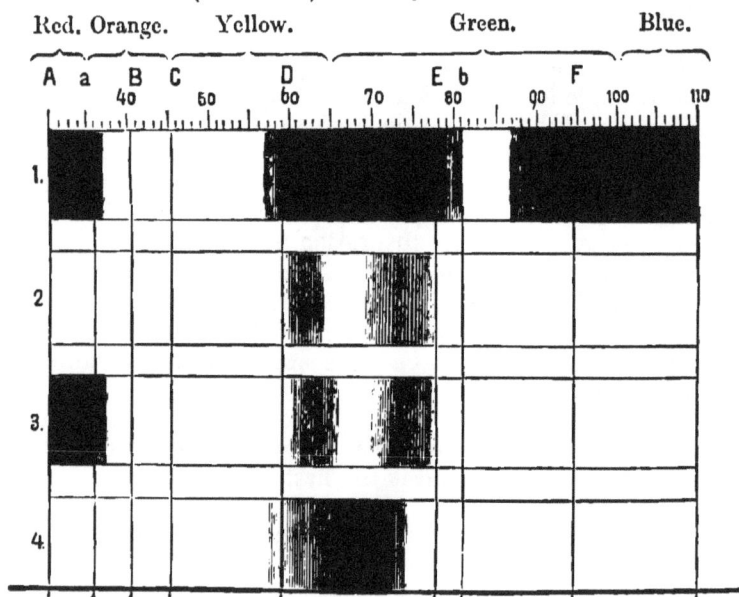

Fig. 4.—Spectra of Hæmoglobin and its compounds.—1, Oxy-hæmoglobin 0·8 per cent.; 2, oxy-hæmoglobin, 0·18 per cent.; 3, carbonic oxide hæmoglobin; 4, reduced hæmoglobin.

9. II. Reduced Hæmoglobin.

(*a.*) To a solution of oxy-hæmoglobin showing two well-defined absorption bands, add a few drops of ammonium sulphide, and warm *gently*, when the solution becomes *purplish* or *claret-coloured*.

(*b.*) Study the spectrum, and note that the two absorption bands of oxy-hæmoglobin are replaced by **one absorption band** between D and E, not so deeply shaded, and with its edges less defined (Fig. 4, 4). By shaking the solution very

vigorously with air, and looking at once, the two bands
may be caused to reappear for a short time. Observe the
absorption of the light at the red and violet ends of the
spectrum according to the strength of the fluid.

(c.) Dilute the solution, and observe that the single band
is not resolved into two bands, but gradually fades and disappears.

(d.) To a similar solution of oxy-hæmoglobin showing
two well-defined bands, add Stokes's fluid, and observe the
single absorption band of reduced hæmoglobin. Shake the
mixture with air and the two bands reappear.

(e.) Use a solution of oxy-hæmoglobin where the two
bands can *just be seen*, and reduce it with either ammonium
sulphide or Stokes's fluid, and note that, perhaps, no absorption band of reduced hæmoglobin is to be seen, or only the
faintest shadow of one.

(f.) Compare the relative strengths of the solution of oxy-
hæmoglobin and reduced hæmoglobin. The latter must be
considerably stronger to give its characteristic spectrum.

Stokes's Fluid.—Make a solution of ferrous sulphate ; to it
add ammonia after the previous addition of sufficient tartaric
acid to prevent precipitation. It is usual to add about three
parts by weight of tartaric acid to two of the iron salt. *It should
be made fresh when required.*

10. III. Carbonic Oxide-Hæmoglobin.—Through a diluted solution of oxy-hæmoglobin, or defibrinated blood, pass a stream of
carbonic oxide—or coal gas—until no more CO is absorbed.
Note the florid cherry-red colour of the blood.

(a.) Dilute the solution in a test-tube and observe its
spectrum, noting that a stronger solution is required than
with HbO_2 to show the absorption bands. **Two absorption
bands** nearly in the same position as those of HbO_2, but
very slightly nearer the violet end (Fig. 4, 3). Make a map
of the spectrum and bands.

(b.) **The bands are not affected by the addition of a re-**
ducing agent—*e.g.*, ammonium sulphide or Stokes's fluid.

3

Add these fluids to two separate test-tubes of the solution of COHb, and observe that the two absorption bands are not affected thereby. There is no difference on shaking the solution with air, as the compound is so very stable.

(c.) To a fresh portion of the solution of carbonic oxide hæmoglobin add a 10 per cent. solution of caustic soda = cinnabar-red colour. Compare this with a solution of oxyhæmoglobin similarly treated. The latter gives a brownish-red mass.

11. IV. Acid-Hæmatin.

(a.) To diluted defibrinated blood add water and about 1 cc. of acetic acid, and warm *gently*, when the mixture becomes brownish owing to the formation of acid-hæmatin.

(b.) Observe the spectrum of (a.), noting **one absorption band** to the red side of D near C (Fig. 5, 5). Observe that there is considerable absorption of the blue end of the spectrum.

(c.) The single band is not affected by the usual reducing agents, ammonium sulphide or Stokes's fluid.

N.B.—If acetic acid alone be used to effect the change, observe that only one absorption band is seen.

12. Acid-Hæmatin in Ethereal Solution.

(a.) To defibrinated blood add ether and a large quantity of strong acetic acid, which makes the mixture brown. Shake vigorously, and a dark-brown ethereal solution of hæmatin is obtained.

(b.) Observe the spectrum of this solution—*four* absorption bands are obtained, one in the red between C and D, corresponding to the watery acid-hæmatin solution; a narrow faint one near D, one between D and E, and a fourth between b and F (Fig. 5, 5). The last three bands are seen only in ethereal solutions, and require to be looked for with care.

13. V. Alkali-hæmatin.

(a.) Take a solution of acid-hæmatin; neutralise it with caustic soda until there is a precipitate of hæmatin; on adding more soda, and heating gently, the precipitate is re-dissolved and alkali-hæmatin is formed.   Or to diluted blood add a drop or two of solution of caustic potash, and warm gently.   The colour changes, and the solution is dichroic.

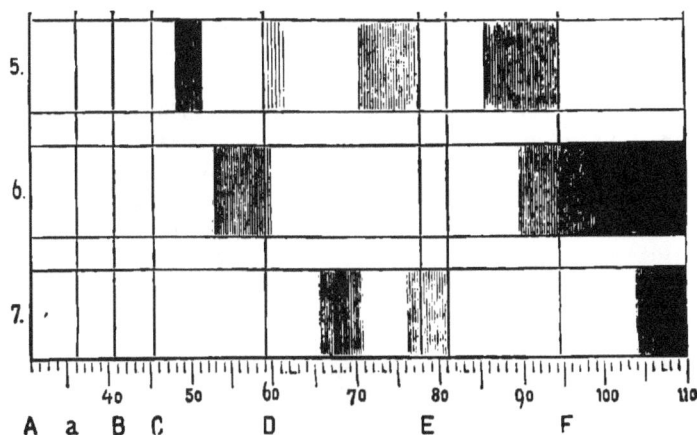

Fig. 5.—Spectra of Derivatives of Hæmoglobin.—5, Hæmatin in alcohol with sulphuric acid ; 6, hæmatin in an alkaline solution ; 7, reduced hæmatin.

(b.) Shake (a.) with air to obtain oxy-alkali-hæmatin. Observe its spectrum, one absorption band just to the *red* side of the D line.   It is much nearer D than that of acid-hæmatin (Fig. 5, 6).   Much of the blue end of the spectrum is cut off.

14. Reduced Alkali-hæmatin or Hæmo-chromogen.

(a.) Add to 13, V. (b.) a drop or two of ammonium sulphide and warm gently = reduced alkali-hæmatin, Stokes's reduced hæmatin, or hæmo-chromogen, and observe its spectrum ; two absorption bands between D and E, as with HbO_2 and HbCO, but they are nearer the violet end.   The first band to the violet side of the D line is well-defined, while the second band still nearer the-violet end (in fact it nearly coincides with the E line) is less defined.   They disappear on shaking vigorously with air, and reappear on standing, provided sufficient ammonium sulphide be added.

15. VI. Methæmoglobin.

(*a*.) To a medium solution of oxy-hæmoglobin, add a few drops of a 1 per cent. solution of potassic permanganate, warm gently, observe the change of colour, and examine it with a spectroscope. If the two bands of oxy-hæmoglobin are still present, allow it to stand for some time and examine again. If they persist, carefully add more permanganate until the two bands disappear. Finally, acidify the solution, and with a spectroscope look for the spectrum of methæmoglobin, viz., one absorption band in the red near C, nearly in the same position, but nearer D than the band of acid-hæmatin; the violet end of the spectrum is much shaded. Three other bands are described in the green, especially in dilute solutions. On adding ammonia to render the solution alkaline, the band in the red disappears, and is replaced by a faint band near D.

(*b*.) To an alkaline solution showing the last described spectrum, add ammonium sulphide or Stokes's fluid. This gives the spectrum of reduced hæmoglobin; and on shaking with air, oxy-hæmoglobin is formed.

(*c*.) To a solution of oxy-hæmoglobin, add a crystal or two of potassic chlorate; dissolve it with the aid of gentle heat; after a short time the spectrum of methæmoglobin is obtained.

(*d*.) **Action of Nitrites.**—To diluted defibrinated ox blood, or preferably that of a dog, add a few drops of an alcoholic solution of amyl nitrite. The blood immediately assumes a chocolate colour.

(*e*.) To another portion of diluted blood add a solution of potassic or sodic nitrite. Observe the chocolate colour.

(*f*.) To portions of (*d*.) and (*e*.) add ammonia, the chocolate gives place to a red colour.

(*g*.) Observe the spectrum of (*d*.) and (*e*.) The band in the red is distinct, and is best seen when the solution is of such a strength that only the red rays are transmitted. On dilution, other bands are seen in the green. Add ammonia, and with the change of colour described in (*f*.), the spectrum changes to that described in (*a*.) Add ammonium sulphide or Stokes's fluid, the spectrum of reduced hæmoglobin appears, and on shaking up with air, the bands of oxy-hæmoglobin appear.

16. VII. Hæmatoporphyrin. (Iron-free Hæmatin.)

(*a*.) Place some concentrated sulphuric acid in a test-tube, add some blood, and examine with the spectroscope. Or examine a solution obtained by dissolving some hæmatin in concentrated sulphuric acid, and filtering through asbestos—when a clear purple-red solution is obtained.

(*b*.) Observe two absorption bands, one close to and on the red side of D, and a second half-way between D and E.

(*c*.) To some of the hæmatin solution (in strong sulphuric acid), add a large excess of water, which throws down part of the hæmatoporphyrin in the form of a brown precipitate, which is more copious if the acid be neutralised with an alkali—*e.g.*, caustic soda. Dissolve some of the brown deposit in caustic soda, and examine it spectroscopically.

(*d*.) The spectrum shows **four absorption bands**; a faint band midway between C and D, another similar one between D and E, but close to D; a third band near E; and a fourth one, darkest of all, occupying the greater part of the space between *b* and F, but nearer the former.

In all cases make drawings of what you see, and compare them with the table of characteristic spectra suspended in the Laboratory.

LESSON V.

WAVE-LENGTHS—DERIVATIVES OF HÆMOGLOBIN—ESTIMATION OF HÆMOGLOBIN.

Spectroscopic Determination of-Wave-Lengths.—Use Zeiss's spectroscope, which is provided with an illuminated scale for this purpose.

1. W.L. of Absorption Bands of Oxy-Hæmoglobin.

(*a.*) Arrange the apparatus as shown in Fig. 6. A is the telescope through which the observer looks and sees the

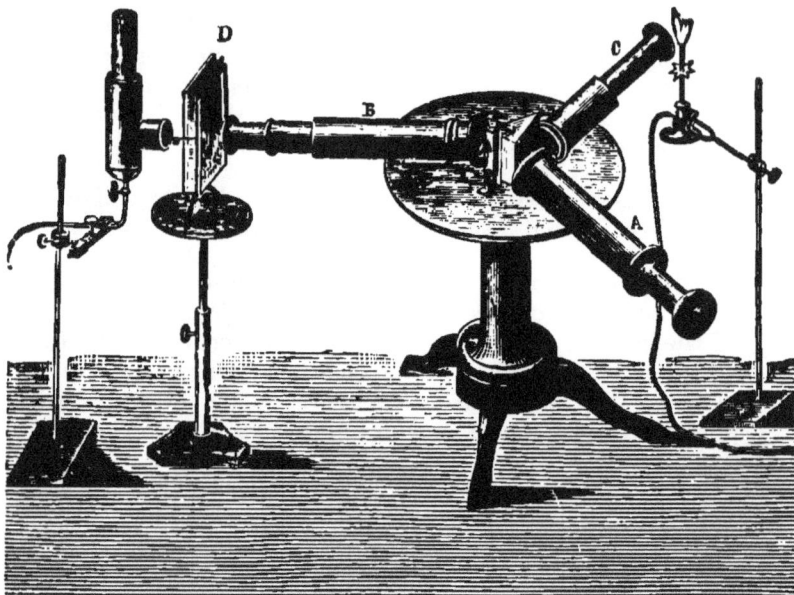

Fig. 6. – Arrangement of the spectroscope for determining wave-lengths.—
A, Telescope ; B, collimator tube ; C, scale tube ; D, hæmatinometer.

spectrum obtained by the light passing through B, and dispersed by the flint-glass prism in the centre of the apparatus. In C is fixed a scale photographed on glass and illuminated by a fan-tailed burner. D is the hæmatinometer containing the dilute blood.

(*b.*) Throw a piece of black velvet cloth over the prism ; light both lamps ; look through B ; adjust the slit in A and the telescope in B, so as to get a good view of the spectrum, and over it the image of the scale. D is supposed not to be in position at first. In a loop of platinum wire burn some common salt in the flame to get the bright yellow sodium line D. Adjust the scale so that this line corresponds to the figures 58·9 on the scale, and fix the spectroscope tubes (A and C) in this position; the scale is now accurately adjusted for all other parts of the spectrum.

"The numbers on the scale indicate wave-lengths expressed in one hundred thousandths of a millimetre, and each division indicates a difference in wave-length equal to one hundred thousandth of a millimetre" (*Gamgee*).

Thus, Frauenhofer's line, D, which corresponds to division 58·9 of the scale, has a wave-length of 589 millionths of a millimetre. The wave-lengths of Frauenhofer's lines are:—
A = 760·4, B = 687·4, C = 656·7, D = 589·4, E = 527·3, F = 486·5.

(*c.*) Using one of the blank maps provided for you (*i.e.*, the maps supplied with Zeiss's spectroscope—the maps correspond to the scale seen in the spectroscope), fill in, in wave-lengths, the position of Frauenhofer's lines, B to F.

(*d.*) Use a dilute solution of blood or hæmoglobin—1 part in 1000 of water is best—and place it in the hæmatinometer, D, which is placed in position between the flame and the spectroscope, as shown in Fig. 6. The distance between the parallel faces of D is 1 cm. The spectrum shows the two absorption bands of oxy-hæmoglobin between D and E. The narrower, sharper, and blacker band near D has its centre corresponding with the W.L. 579, and it may conveniently be expressed by the letter α of the oxy-hæmoglobin spectrum (*Gamgee*).

The other absorption band near E, and conveniently designated β, is broader, not so dark, and has less sharply defined edges than α. Its centre corresponds to the W.L. 553·8. Notice that the other parts of the spectrum are seen, there being only slight cutting off of the red, and a slightly greater absorption of the violet end.

(*e.*) Work with a stronger solution of blood, and observe how the two bands become fused into one, while more and more of the red and violet ends of the spectrum are absorbed as the solution is made stronger, until finally only a little red light is transmitted.

2. W.L. of Absorption Band of Reduced Hb.

(*a.*) Adjust the apparatus as before, but reduce the oxy-hæmoglobin solution with Stokes's fluid—noticing the change of the colour to that of purplish or claret—until a solution is obtained, which gives the single characteristic absorption band of reduced Hb. This is usually obtained with a solution of Hb of about 0·2 per cent.

(*b.*) Observe the single absorption band less deeply shaded, and with less defined edges between D and E, conveniently designated by the letter γ. It extends between W.L. 595 and 538, and is not quite intermediate between D and E; is blackest opposite W.L. 550, so that it lies nearer D than E (*Gamgee*). Both ends of the spectrum are more absorbed than with a solution of oxy-hæmoglobin of the same strength. On further dilution of the solution, the band does not resolve itself into two bands, but simply diminishes in width and intensity.

3. W.L. of the Spectrum of Carbonic Oxide Hæmoglobin.

(*a.*) Use a dilute solution of carbonic oxide-hæmoglobin of such a strength as to give the two characteristic absorption bands.

(*b.*) Observe the two bands, α and β, like those of Hb-O$_2$, but both are *very slightly* more towards the violet end of the spectrum. α extends from about W.L. 587 to 564, and β from 547 to 529 (*Gamgee*).

(*c.*) No reduction is obtained by reducing agents.

4. Preparation of Hæmatin.

(*a.*) Make defibrinated blood into a paste with potassic carbonate. Dry the paste on a water-bath. Place some of the paste in a flask, add 4 volumes of alcohol, and boil on a water-bath. Filter, and an alkaline brown solution of hæmatin is obtained. Re-extract the residue several times with boiling alcohol, and mix the alcoholic extracts. The solution is dichroic.

(*b.*) Acidify the alkaline filtrate of (*a.*) with dilute sulphuric acid, filter, and keep the filtrate. Observe the spectrum of acid-hæmatin in the filtrate (Fig. 5, 5).

(*c.*) Add excess of ammonia to the acid filtrate of (*b.*), and filter off the precipitate, keep the filtrate, and observe that it is dichroic. Observe the spectrum of alkali-hæmatin in the filtrate (Fig. 5, 6).

(d.) Evaporate the filtrate from (c.) to dryness on a water-bath. Extract the residue with boiling water. The black residue is washed on a filter with distilled water, alcohol, and ether, and dried in a hot chamber at 120° C. This is nearly pure hæmatin.

(e.) It is convenient to keep in stock hæmatin prepared as follows :—Extract defibrinated blood or blood-clot (ox or sheep) with rectified spirit containing pure sulphuric acid (1 : 20). Filter, the solution gives the spectrum of acid-hæmatin. Add an equal volume of water and then chloroform. The chloroform becomes brown, and there is a precipitate of proteids. Separate the chloroform extract, wash it with water to remove the acid. Separate the chloroform, and allow it to evaporate. The dark brown residue is impure hæmatin. When dissolved in alcohol and caustic soda it gives the spectrum of alkali-hæmatin, and on adding ammonium sulphide that of hæmochromogen. If it is dissolved in H_2SO_4, and filtered through asbestos, the red filtrate gives the spectrum of hæmatoporphyrin (*MacMunn*).

5. Hæmin Crystals.—Place some powdered dried blood on a glass slide, add a crystal of sodium chloride, and a few drops of *glacial* acetic acid. Cover with a cover-glass, and heat *carefully* over a flame until bubbles of gas are given off. After cooling, brown or black rhombic crystals of hæmatin are to be seen with a microscope (Fig. 7).

6. Detection of Blood Stains.—Use a piece of rag stained with blood.

(a.) Moisten a part of the stain with glycerin, and after a time express the liquor, and observe it microscopically for blood-corpuscles.

Fig. 7.—Hæmin Crystals prepared from traces of blood.

(b.) Tie a small piece of the stained cloth to a thread, place the cloth in a test-tube with a few drops of distilled water, and leave it until the colouring-matter is extracted.

Withdraw the cloth by means of the thread. Observe the coloured fluid spectroscopically.

(c.) Boil some of the extract with hydrochloric acid, and add potassic ferrocyanide; a blue colour indicates the presence of iron.

(d.) Use the stain for the hæmin test, doing the test in a watch-glass (Lesson V., 5).

AMOUNT OF HÆMOGLOBIN IN BLOOD.

7. Colorimetric Method (Hoppe-Seyler's Method).—A standard solution of pure hæmoglobin diluted to a known strength is used, and with this is compared the tint of the blood diluted with a known volume of distilled water.

(a.) The demonstrator will prepare a standard solution of hæmoglobin of known strength.

(b.) Spread a sheet of white paper on a table in a good light opposite a window, and on it place two hæmatinometers side by side (Fig. 6, D). See that they are watertight. If not, anoint the edges of the glass plates with vaseline to make them water-tight.

(c.) Take 10 cc. of the standard solution of hæmoglobin and dilute it with 50 cc. of water, and place it in one of the hæmatinometers.

(d.) Weigh 5 grammes of the blood to be investigated, and dilute it with water exactly to 100 cc.

(e.) Place 10 cc. of this deeper tinted blood (d.) into the second hæmatinometer.

(f.) Fill an accurately graduated burette with distilled water, place it over the second hæmatinometer (e.), and dilute the blood in it until it has precisely the same tint as the standard solution in the other hæmatinometer. Note the amount of water added. The two solutions must now contain the same amount of hæmoglobin.

Example (*Hoppe-Seyler*).—20·186 grms. of defibrinated blood were diluted with water to 400 cc. To the 10 cc. of this placed in a hæmatinometer, 38 cc. of water had to be added to obtain the same tint as that of the standard solution, so that the volume of water which would require to be added to dilute the whole 400 cc. would be 1,520 cc., thus—

$$10 \; : \; 400 \; : : \; 38 \; : \; x$$
$$x = 1,520 \text{ cc.}$$

By adding 1,520 cc. of distilled water to the 400 cc. of blood solution, we get 1,920 cc. of the same tint or degree of dilution as the standard solution.

The standard solution on analysis was found to contain 0·145 grms. of hæmoglobin in 100 cc., so that the total amount of the hæmoglobin in the diluted blood is found, thus—

$$100 \; : \; 1,920 \; : : \; 0·145 \; : \; x$$
$$x = 2·784 \text{ grms.}$$

Since, however, this amount of hæmoglobin was obtained from 20·186 grms. of the original blood, the amount in 100 parts will be found, as follows:—

$$20·186 \; : \; 100 \; : : \; 2·784 \; : \; x$$
$$x = 13·79 \text{ grms. per cent.}$$

8. **The Hæmoglobinometer** of Gowers is used for the clinical estimation of hæmoglobin (Fig. 8). The tint of the dilution of a given volume of blood with distilled water is taken as the index of the amount of hæmoglobin. The colour of a dilution of average normal blood (one hundred times) is taken as the standard. The quantity of hæmoglobin is indicated by the amount of distilled water needed to obtain the tint with the same volume of blood under examination as was taken of the standard. On account of the instability of a standard dilution of blood, tinted glycerin-jelly is employed instead. The apparatus consists of two glass tubes of exactly the same size. One contains (D) a standard of the tint of a dilution of 20 c.mm. of blood, in 2 cc. of water (1 in 100). The second tube (C) is graduated, 100° = 2 c. (100 times 20 c.mm.) The 20 c.mm. of blood are measured by a capillary pipette (B).

(*a*) Place a few drops of distilled water in the bottom of the graduated tube (C).

(*b*.) Puncture the skin at the root of the nail with the shielded lancet (F), and with the pipette (B) suck up 20 c.mm. of the blood, and eject it into the distilled water, and rapidly mix them.

(*c*.) Distilled water is then added drop by drop (from the pipette stopper of a bottle (A) supplied for that purpose)

until the tint of the dilution is the same as that of the
standard.   The amount of water which has been added (*i.e.*,
the degree of dilution) indicates the amount of hæmoglobin.

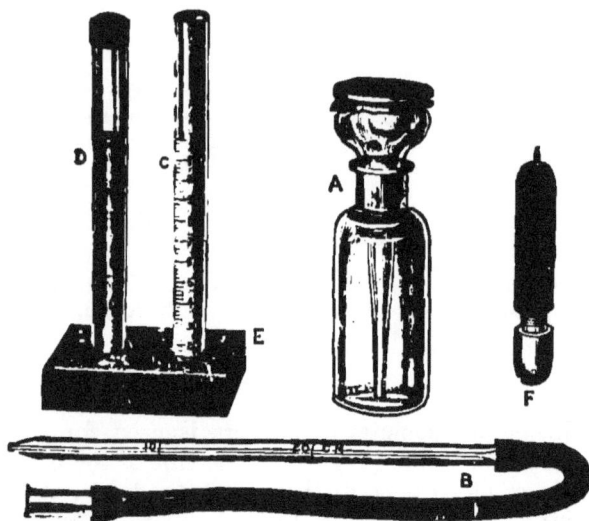

Fig. 8.—A, Pipette bottle for distilled water ; B, capillary pipette ; C,
   graduated tube ; D, tube with standard dilution ; F, lancet for
   pricking the finger.

" Since average normal blood yields the tint of the standard at
100° of dilution, the number of degrees of dilution necessary to
obtain the same tint with a given specimen of blood is the percentage proportion of the hæmoglobin contained in it, compared
to the normal.   For instance, the 20 c.mm. of blood from a
patient with anæmia gave the standard tint of 30° of dilution.
Hence it contained only 30 per cent. of the normal quantity of
hæmoglobin.   By ascertaining with the hæmacytometer the cor-
puscular richness of the blood, we are able to compare the two.
A fraction, of which the numerator is the percentage of hæmoglobin, and the denominator the percentage of corpuscles, gives
at once the average value per corpuscle.   Thus the blood mentioned above containing 30 per cent. of hæmoglobin, contained
60 per cent. of corpuscles ; hence the average value of each cor-
puscle was $\frac{30}{60}$ or $\frac{1}{2}$ of the normal.   Variations in the amount of
hæmoglobin may be recorded on the same chart as that employed
for the corpuscles."

" In using the instrument, the tint may be estimated by holding the tubes between the eye and the window, or by placing a piece of white paper behind the tubes ; the former is perhaps the best. In practice it will be found that, during 6 or 8 degrees of dilution, it is difficult to distinguish a difference between the tint of the tubes. It is therefore necessary to note the degree at which the colour of the dilution ceases to be deeper than the standard, and also that at which it is distinctly paler. The degree midway between these two will represent the hæmoglobin percentage."

9. **Fleischl's Hæmometer.**—This apparatus (Fig. 9) consists of

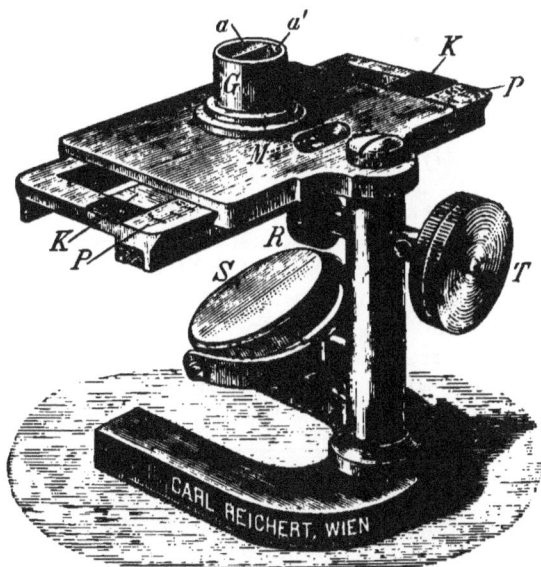

Fig. 9.—Fleischl's Hæmometer.

a horse-shoe stand with a pillar bearing a reflecting surface (S) and a platform. Under the table or platform is a slot carrying a glass wedge stained red (K), and moved by a wheel (R). On the platform (M) is a small cylindrical vessel divided into two compartments (a and a') by a vertical septum. In one compartment is placed pure water, and in the other the blood to be investigated. A scale (P) on the slot of the instrument enables one to read off directly the percentage of hæmoglobin.

(a.) Fill with a pipette the compartment (a') over the wedge with distilled water, and see that the surface of the water is quite level with the top of the cylinder. Fill the other compartment (*a*), that for the blood, about one-quarter with distilled water.

(*b*.) Prick the finger as in 8 with the instrument supplied for the purpose. Fill the short automatic capillary pipette tube with blood. Its capacity is 6·5 c.mm. In filling the tube, hold it horizontally. See that no blood adheres to the surface of the tube. This can be done by having the pipette slightly greasy on the outer surface.

(*c*.) Dissolve the blood obtained in (*b*.) in the water of the blood-compartment (*a'*), washing out every trace of blood from the pipette. Mix the blood and water thoroughly. Clean the pipette. Then fill the blood-compartment exactly to the surface with distilled water, seeing that its surface also is perfectly level.

(*d*.) Arrange a lamp in front of the reflector (S)—which is white, and with a smooth matt surface made of plaster-of-Paris—so as to throw a beam of light vertically through both compartments. Look down vertically upon both compartments, and move the wedge of glass by the milled head (T) until the colour in the two compartments is identical. Read off the scale, which is so constructed as to give the percentage of hæmoglobin.

CHEMISTRY OF DIGESTION.

LESSON VI.

SALIVARY DIGESTION.

1. To obtain mixed Saliva.—Rinse out the mouth with water. Inhale the vapour of ether, glacial acetic acid, or even cold air through the mouth, which causes a reflex secretion of saliva. In doing so, so curve the tongue and place its tip behind the incisor teeth of the upper jaw. In a test-glass collect the saliva with as few air-bubbles as possible. If it be turbid or contain much froth, filter it through a small filter.

2. I. Microscopic Examination.—With a high power observe the presence of (1) squamous epithelium, (2) salivary corpuscles, (3) perhaps *débris* of food, and (4) possibly air-bubbles.

II. Physical and Chemical Characters.

(*a.*) Observe its appearance—either transparent or translucent—and that when poured from one vessel to another it is glairy and more or less sticky. On standing, a white deposit is apt to form.

(*b.*) Test its *reaction*, neutral or alkaline.

(*c.*) Place a little mixed saliva in a test-tube, add dilute acetic acid = a precipitate of **mucin**. Filter.

(*d*). With the filtrate from (*c.*), test for traces of **proteids** (albumin and globulin) with the xanthoproteic reaction (Lesson I., 1, *a*), or by the addition of potassium ferrocyanide.

(*e.*) To a few drops of saliva in a porcelain vessel, add a few drops of *dilute* ferric chloride = a red coloration, due to **potassic sulpho-cyanide.** The colour is discharged by mercuric chloride. Meconic acid yields a similar colour, but it is not discharged by mercuric chloride. The sulpho-cyanide is present only in parotid saliva, and is generally present in mixed saliva.

(*f.*) Test a very dilute solution of potassic sulpho-cyanide to compare with (*e.*)

(*g.*) The **salts** are tested for in the usual way (see " Urine "). Test for chlorides (HNO_3, and $AgNO_3$), carbonates (acetic acid), and sulphates (barium nitrate and nitric acid).

3. Digestive Action.

Starch Solution.—Place 1 grm. of starch in a mortar, add a few cc. of cold water, and mix well with the starch. Add 200 cc. of boiling water, stirring all the while. Boil the fluid in a flask for a few minutes. This gives a half per cent. solution. Do the tests for starch already described (Lesson II., 2), and especially satisfy yourself that no glucose or reducing sugar is present.

Action of Saliva on Starch.

(*a.*) Dilute the saliva with 5 volumes of water. Label three test-tubes A, B, and C. In A place starch mucilage, in B saliva, and in C 1 volume of saliva and 3 volumes of starch mucilage. Plug all three with cotton-wool, and place them in a water-bath at 40° C., and leave them there for ten minutes. Test for a reducing sugar in portions of all three, by means of Fehling's solution. A and B give no evidence of sugar, while C reduces the Fehling, giving a yellow or red deposit of cuprous oxide. Therefore, starch is converted into a reducing sugar by the saliva. This is done by the ferment **ptyalin** contained in it.

(*b.*) Test a portion of C with solution of iodine ; no blue colour is obtained, as all the starch has disappeared, being converted into a reducing sugar or **maltose.**

(*c.*) Make a *thick* starch mucilage, place some in test-tubes labelled A and B. Keep A for comparison, and to B add saliva, and expose both to 40° C. Notice that A is un-affected, while B soon becomes fluid—within two minutes—

and loses its opalescence; this liquefaction is a process quite antecedent to the saccharifying process which follows.

4. Stages between Starch and Maltose.—Mix some starch and saliva in a test-tube as in 3 (*a*.) C, and place it in a water-bath at 40° C. At intervals of a minute, test small portions with iodine. Do this by taking out a drop of the liquid by means of a glass rod. Place the drop on a white porcelain plate, and by means of another glass rod add a drop of iodine solution.

Note the following stages—At first there is pure blue with iodine, later a deep violet, showing the presence of **erythrodextrin**, the violet resulting from a mixture of the red produced by the dextrin and the blue of the starch. Then the blue reaction entirely disappears, and a reddish-brown colour, due to erythro-dextrin alone is obtained. After this the reaction becomes yellowish-brown, and finally there is no reaction with iodine at all, **achroo-dextrin** being formed, along with a reducing sugar or **maltose**. The latter goes on forming after iodine has ceased to react with the fluid, and its presence is easily ascertained by Fehling's solution.

5. Effect of Different Conditions on Salivary Digestion.

(*a*.) Label three test-tubes A, B, and C. Into A place some saliva and boil it, add some starch mucilage. In B and C place starch mucilage and saliva, to B add a few drops of hydrochloric acid, and to C caustic potash. Place all three at 40° C. on a water-bath, and after a time test them for sugar by Fehling's solution. No sugar is formed—in A because the ferment was destroyed by boiling, and in B and C because strong acids and alkalies arrest the action of ptyalin on starch.

(*b*.) If a test-tube containing starch mucilage and saliva be prepared as in 3 (*a*.) C, and placed in a freezing-mixture, the conversion of starch into a reducing sugar is arrested; but the ferment is not killed, for on placing the test-tube at 40° C. the conversion is rapidly effected.

(*c*.) Mix some **raw starch** with the saliva and expose it to 40° C. Test it after half an hour or longer, when no sugar will be found.

6. Starch is a Colloid, but Sugar dialyses.

(*a.*) Place in a short piece of the sausage parchment tube, already referred to (Lesson I. 1, *h*), 20 cc. of starch mucilage, label it A, and into another, some starch mucilage with saliva, label it B. Suspend A and B in distilled water in separate vessels.

(*b.*) After some hours test the diffusate in the distilled water. No starch will be found in the diffusate of either A or B, but sugar will be found in that of B, proving that sugar dialyses, while starch does not. Hence the necessity of starch being converted into a readily diffusible body during digestion.

7. Action of Malt-Extract on Starch.

(*a.*) Rub up 10 grms. of starch with 30 cc. of distilled water in a mortar, add 200 cc. of boiling water, and make a strong starch mucilage.

(*b.*) Powder 5 grms. of pale *low dried* malt, and extract it for half an hour with 30 cc. of distilled water, and filter. Keep the filtrate.

(*c.*) Place the starch paste of (*a.*) in a flask, and cool to 60° C., add the extract of (*b.*), and place the flask in a water-bath at 60° C.

(*d.*) Observe that the paste soon becomes fluid, owing to the formation of soluble starch, and if it be tested from time to time (as directed in **4**), it gives successively the tests for starch and erythro-dextrin. Continue to digest it until no colour is obtained with iodine.

(*e.*) Take a portion and precipitate it with alcohol = **achroo-dextrin**.

(*f.*) Boil the remainder of the fluid, cool, filter, and evaporate the filtrate to 20 cc. Add 6 volumes of 90 per cent. spirit to precipitate the dextrin, boil, filter, and concentrate to dryness on a water-bath, and dissolve the residue in distilled water. The solution is **maltose** ($C_{12}H_{22}O_{11} + H_2O$). If the alcoholic solution be exposed to air, crystals of maltose are formed.

8. Compare the Reducing Power of Maltose and Dextrose.

(*a.*) With Fehling's solution estimate the reducing power of the solution obtained in 7 (*f.*) (See "Urine.")

(*b.*) Boil in a flask for half an hour 50 cc. of the solution of maltose with 5 cc. of hydrochloric acid. Neutralise with caustic soda, and make up the volume, which has been reduced by the boiling, to 50 cc., and determine by Fehling's solution the reducing power. The acid has converted the maltose into dextrose, and the ratio of the former estimation (*a.*) to the present one should be 66 to 100.

(*c.*) A solution of pure dextrose treated as in (*b.*) is not affected in its reducing power.

Saliva has practically the same effect on starch as malt-extract, and may be used instead of the latter.

LESSON VII.

GASTRIC DIGESTION.

1. Preparation of Artificial Gastric Juice.

(*a.*) Take a part of the cardiac end of the pig's stomach provided for you, which has been previously opened and washed rapidly in cold water. Spread it, mucous surface upwards, on the convex surface of an inverted capsule. Scrape the mucous surface firmly with the handle of a scalpel, and rub up the scrapings in a mortar with fine sand. Add water, and rub up the whole vigorously for some time, and filter. The filtrate is an artificial gastric juice.

(*b.*) From another portion of the cardiac end of a pig's stomach detach the mucous membrane in shreds, dry them between folds of blotting-paper, place them in a bottle, and cover them with strong glycerin, letting them stand for eight days. The glycerin dissolves out the pepsin, and on filtering, a glycerin extract with high digestive properties is obtained (**v. Wittich's Method**).

(*c.*) Instead of (*a.*) and (*b.*) it is convenient to use Benger's Liquor pepticus, or the pepsin preparation of Burroughs, Wellcome & Co.

All the above artificial juices, when added to hydrochloric acid of the proper strength, have high digestive powers.

2. Both Hydrochloric Acid and Pepsin are required for Gastric Digestion.

(*a.*) Take three beakers or large test-tubes, label them A, B, C. Fill A two-thirds full of hydrochloric acid 0·2 per cent., put into B some water and a few drops of glycerin extract of pepsin, or powdered pepsin, and fill C two-thirds full with 0·2 per cent. hydrochloric acid, and a few drops of glycerin extract of pepsin. Put into all three a flake of well-washed fibrin, and place them all in a water-bath at 40° C. for half an hour.

(*b.*) Examine them. In A, the fibrin is swollen up ; in B, unchanged ; while in C, it has disappeared, having first become swollen up and clear, and finally completely dissolved, being converted into peptones. Therefore, both acid and ferment are required for gastric digestion.

3. To Prepare Hydrochloric Acid of 0·2 per cent.—Add 6·5 cc. of ordinary commercial hydrochloric acid to 1 litre of distilled water, and shake together.

4. Products of Peptic Digestion and its Conditions.

(*a.*) Take three large test-tubes, labelled A, B, C, and fill each one half full with hydrochloric acid 0·2 per cent. Add to each 10 drops of a glycerin extract of pepsin. Boil B, and make C faintly alkaline with sodic carbonate. The alkalinity may be noted by adding previously some neutral litmus solution. Add to each an equal amount—a few shreds of well-washed fibrin—which has been previously steeped for some time in 0·2 per cent. hydrochloric acid, so that it is swollen up and transparent. Keep the tubes in a water-bath at 40° C. for an hour, and examine them at intervals of twenty minutes.

(*b.*) After twenty minutes A begins to be turbid, and the fibrin is dissolving. In B and C there is no change. After forty minutes A is turbid, and the fibrin is dissolved. In B and C no change. At the end of an hour, filter A and part of B and C. Keep the filtrates.

(c.) Carefully neutralise the filtrate of A with dilute caustic soda. The filtrate becomes turbid and gives a precipitate of **parapeptones** (antialbumose and hemialbumose). Filter off this precipitate, dissolve it in 0·2 per cent. hydrochloric acid. It gives proteid reactions.

(d.) With a solution of **parapeptones** (hemialbumose) repeat the ordinary reactions for proteids. **Hemialbumose** is soluble in water, and gives all the ordinary proteid reactions. It is precipitated by nitric acid in the cold, but the precipitate is redissolved with the aid of heat.

(e.) Test the filtrate of (c.) for **peptones.** Repeat all the tests for peptones (Lesson I., 10, VI.) Note that hemialbumose gives the ordinary proteid reactions. Note also the differences between peptones and hemialbumose. Hemialbumose is precipitated by acetic acid and ferrocyanide of potassium ; by acetic acid and a saturated solution of sodic sulphate; and by metaphosphoric acid: which peptones are not. Like peptones, it is soluble in water.

(f.) Neutralise part of the filtrates of B and C. They give no precipitate, nor do they give the reactions for peptones. In B the ferment pepsin was destroyed by boiling, while in C the ferment cannot act in an alkaline medium.

(g.) If to the remainder of C acid be added, and it be placed again at 40° C., digestion takes place, so that neutralisation has not destroyed the activity of the ferment.

5. To prepare Hemialbumose and Gastric Peptones in Quantity.

(a.) Place 10 grms. of fresh, well-washed, expressed fibrin in a porcelain capsule, cover it with 300 cc. of 0·2 per cent. hydrochloric acid, and keep the whole at 40° C. in a water-bath until the whole of the fibrin is so swollen up as to become converted into a perfectly clear, jelly-like mass, and it becomes so thick that a glass rod is supported erect in it.

(b.) Add 1 or 2 cc. of glycerin pepsin extract, and stir

the mass.   Within a few minutes the whole becomes fluid.

(c.) After a short time—fifteen to twenty minutes—before the peptonisation is complete, filter and exactly neutralise the filtrate with ammonia, which precipitates the **antialbumose** and **hemialbumose**.   Dissolve these in a 5 per cent. solution of sodic chloride.   To isolate the hemialbumose, precipitate it with nitric acid, or dialyse the salt solution of it in a parchment-paper tube arranged in Kühne's dialyser (Fig. 10).   The greater part of the hemialbumose is thrown down in flocculi on the parchment tube.

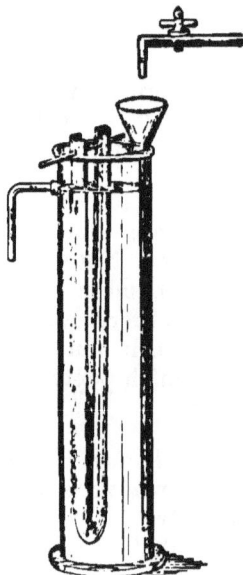

(d.) The filtrate after neutralisation is evaporated, and yields **peptones**, which can be precipitated by alcohol.

6. Action of Gastric Juice on Milk.

(a.) Place 5 cc. of fresh milk in a test-tube, and add to it a few drops of a *neutral* artificial gastric juice.   Mix and keep at 40° C.   In a short time the milk curdles, so that the tube can be inverted without the **curd** falling out.   By-and-by **whey** is squeezed out of the clot.   The curdling of milk by the rennet ferment present in the gastric juice is quite different from that produced by the "souring of milk," or by the precipitation of casein by acids.   Here the casein (carrying with it most of the fats) is precipitated in a neutral fluid.

Fig. 10.
Kühne's Dialyser.

(b.) To the test-tube add 5 cc. of 0·4 per cent. hydrochloric acid, and keep at 40° C. for two hours.   The pepsin in the presence of the acid digests the casein, gradually dissolving it, forming a straw-yellow coloured fluid containing peptones.   The "peptonised milk" has a peculiar odour and bitter taste.

(*c.*) To 5 cc. of milk in a test-tube add a few drops of Benger's liquor pepticus, and place in a water-bath. Observe how the casein first clots, and is then partially dissolved to form a yellowish coloured fluid, with a bitter taste and peculiar odour. There generally remains a very considerable clot of casein; and, in fact, the gastric digestion of milk is slow, especially if compared with its tryptic digestion (Lesson VIII., 9). Test the fluid for peptones with the biuret test, and observe the beautiful light pink colour obtained. The bitter taste renders milk "peptonised" by gastric juice unsuitable for feeding purposes.

7. Action of Rennet on Milk.

(*a.*) Place some milk in a test-tube, label it A, add a drop or two of rennet, shake it up, and place the tube in a water-bath at 40° C. Observe that the milk becomes solid in a few minutes, forming a curd, and by-and-by the curd of casein contracts and squeezes out a fluid—the whey.

(*b.*) Repeat the same experiment, but previously boil the rennet. No such result is obtained as in (*a.*)

8. Comparison of Mineral and Organic Acids.

(*a.*) Take two test-tubes, A and B. Place in A 10 cc. of a 0·2 per cent. solution of hydrochloric acid, and in B 10 cc. of a 5 per cent. solution of acetic acid. To both add a few drops of o-o-Tropæolin dissolved in alcohol. The very dilute mineral acid in A renders it rose-pink, while the far stronger organic acid does not affect its colour.

(*b.*) Repeat (*a.*), but add to the acids a dilute solution of methyl-violet, and note the change of colour produced by the mineral acid. It becomes blue.

(*c.*) Repeat (*a.*) with the same acids, but use threads stained with congo-red, and observe the change of colour to blue produced by the hydrochloric acid.

(*d.*) Instead of threads stained with congo-red, use papers similarly stained, or a watery solution of congo-red.

(*e.*) For lactic acid. Prepare a fresh solution by mixing 10 cc. of a 4 per cent. solution of carbolic acid, with 20 cc.

of distilled water, and 1 drop of liquor ferri perchloridi. The blue solution thus obtained is changed to *yellow* by lactic acid, while it is not affected by 0·2 per cent. HCl (Uffelmann's Reaction).

These reactions for a mineral acid are specially to be noted, as they are sometimes used clinically for ascertaining the presence or absence of hydrochloric acid—*e.g.*, in a vomit. This acid is almost invariably absent from the gastric juice in cancer of the stomach. It is to be noted, however, that the presence of peptones interferes with the delicacy of these reactions.

LESSON VIII.

PANCREATIC DIGESTION.

1. Preparation of Artificial Pancreatic Juice.

(*a*.) Use part of the pancreas of an ox twenty-four hours after the animal was killed. Mince a portion of the pancreas, rub it up with well-washed fine sand in a mortar, and digest it with cold water, stirring vigorously. After a time strain through muslin, and then filter through paper. The filtrate has digestive properties chiefly upon starch. Instead of water a more potent solution is obtained by digesting the pancreas at 40° C. for some hours with a 2 per cent. solution of sodic carbonate. To prevent the putrefactive changes which are so apt to occur in all pancreatic fluids, add a little 10 per cent. alcoholic solution of thymol.

(*b*.) Make a glycerin extract of the pancreas in the same way as described for the stomach (Lesson VII., 1, *b*.) Before putting it in glycerin it is well to place it for two days in absolute alcohol to remove all the water. This extract acts on starch and proteids.

(*c*.) For most experiments it is more convenient to use the excellent pancreatic extracts now supplied by Mr. Benger, of Manchester, as "Liquor Pancreaticus," or those of Messrs. Savory & Moore, or Burroughs, Wellcome & Co.

(*d.*) Weigh the pancreas taken from a dog just killed, rub it up with sand in a mortar, and add 1 cc. of a 1 per cent. solution of acetic acid for every gramme of pancreas. Mix thoroughly, and after a quarter of an hour add 10 cc. of glycerin for every gramme of pancreas. After five days filter off the glycerin extract. The acetic acid is added to convert the unconverted " zymogen " into trypsin.

(*e.*) **Kühne's Dry Pancreas Powder.**—This is obtained by thoroughly extracting a pancreas with alcohol and ether, and drying the residue. It is better to purchase the preparation. Extract the dry pancreas powder with five parts of a 1 per cent. solution of salicylic acid, and keep it about 40° C. for four or five hours. Filter, and use the filtrate as a glycerin or other extract would be used. It has only proteolytic properties. I find this extract acts much more energetically than those prepared in other ways.

2. I. Action on Starch.

(*a.*) Take thick starch mucilage in a test-tube or beaker, add glycerin extract of pancreas or liquor pancreaticus, and place it in a water-bath at 40° C. Almost immediately the starch paste becomes fluid, loses its opalescence, and becomes clear. Within a few minutes much of the starch is converted into a reducing sugar or **maltose.**

(*b.*) Test for sugar (Lesson II., **6, IV.,** *b.*, *c.*)

3. The same **conditions** obtain as for saliva (Lesson VI., **5**).

4. II. Proteolytic Action due to Trypsin, and its Conditions.

(*a.*) Take three test-tubes, labelled A, B, and C, fill each half full with 1 per cent. solution of sodic carbonate, and place 5 drops of glycerin pancreatic extract, or liquor pancreaticus in each. Boil B, and make C acid with dilute hydrochloric acid. Place in each tube an equal amount of well-washed fibrin, plug the tubes with cotton-wool, and place all in a water-bath at 40° C.

(*b.*) Examine them from time to time. At the end of one hour or so, there is no change in B and C, while in A the fibrin is gradually being *eroded,* and finally disappears, but it does not swell up, the solution at the same time

becoming slightly turbid.   After two hours, still no change
is observable in B and C.

(c.) Filter A, and carefully neutralise the filtrate with
very dilute hydrochloric or acetic acid = a precipitate of
hemialbumose.   Filter off the precipitate, and on testing the
filtrate, **peptones** are found.

(d.) Filter B and C, and carefully neutralise the filtrates.
They give no precipitate.   No peptones are found.

(e.) Test the proteolytic power of an extract of Kühne's
"pancreas powder" (Lesson VIII., 1, e.)

5. Products other than Peptones.   Leucin and Tyrosin (Indol).

(a.) Place 300 cc. of a 1 per cent. solution of sodic
carbonate in a flask, add 5 grammes of boiled fibrin, 5 cc.
of glycerin extract of pancreas, and a few drops of an alcoholic
solution of thymol.   Keep all at 38° C. on a water-bath
for six to ten hours.

(b.) After six hours take a portion of the mixture, filter,
and to the filtrate cautiously add dilute acetic acid to precipitate any hemialbumose that may be present in it.   Filter
and evaporate the filtrate to a small bulk, and precipitate
the peptones by a considerable volume of alcohol.   Filter to
remove the peptones, and evaporate the alcoholic filtrate to
a small bulk, and set it aside, when leucin separates first, and
crystals of tyrosin afterwards.   Keep them for microscopic
examination.

(c.) A much better method of obtaining **leucin and tyrosin**
is to digest, at 40° C., for five or six hours, equal parts of
fresh moist fibrin and ox-pancreas with a sufficient quantity
of thymolised water.   Boil part of the liquid, and evaporate
a small quantity of it, or merely place a drop on a glass
slide and allow it to evaporate, when beautiful microscopic
crystals of leucin and tyrosin are obtained.   Continue the
digestive process of the remainder of the liquid for a few
hours, until the mixture emits a very disagreeable odour.
This fluid gives the chlorine and indol reaction splendidly.

(d.) Examine the crystals of **leucin and tyrosin** microscopically.   The former occurs as brown balls, often with radiating lines, not unlike fat, but much less refractive, and the

latter consists of long white shining needles arranged in a stellate manner, or somewhat felted (see " Urine," Fig. 32).

(*e.*) **Test for Tyrosin** (Hofmann).—Dissolve some crystals by boiling them in water, add Millon's reagent, and boil, which gives a rosy-red colour.

(*f.*) Test a solution of tyrosin obtained by the prolonged boiling of horn shavings and sulphuric acid, with Millon's reagent, as in (*e.*)

(*g.*) **Indol.**—The remainder of the original digestive fluid after digestion for ten hours or longer, emits an intensely disagreeable odour, due to indol, whose presence is ascertained by warming the liquid, and adding first a drop or two of dilute sulphuric acid to some of the filtered liquid, and then a *very dilute nitrite* solution. A red colour indicates the presence of indol. This test is very readily obtained with the products of digestion by Kühne's dry pancreas (Lesson VIII., 1, *e*). One must be careful to regulate the strength of the acid.

(*h.*) Acidify strongly with hydrochloric acid a small quantity of the highly offensive fluid, and place in it a shaving of wood or a wooden match with its head removed and soaked in strong hydrochloric acid. The match is coloured a beautiful red, sometimes even an intense red. The match can be dried, and it keeps its colour for a long time, although the colour darkens and becomes somewhat duskier on drying.

(*i.*) **Chlorine Reaction.**—Add to some of the digestive fluid (*g*, or preferably *c*), drop by drop, chlorine water, it strikes a beautiful rosy-red tint. Or add very dilute bromine water (1 to 2 drops to 60 cc. water), the fluid first becomes pale red, then violet, and ultimately deep violet (Kühne).

III. The Action on Fats is twofold.

6. A. Emulsification.

(*a.*) Rub up in a mortar which has been warmed in warm water, a little olive oil or melted lard, and some pieces of fresh pancreas. A creamy, persistent emulsion is formed. Examine the emulsion under the microscope. Or use a watery extract of the fresh pancreas, and do likewise; but in this case the result will not be nearly so satisfactory.

(*b.*) Rub up oil as in (*a.*); but this time use an extract of the fresh pancreas made with 1 per cent. sodic carbonate. A very perfect emulsion is obtained, even if the sodic carbonate extract is boiled beforehand. This shows that its emulsifying power does not depend on a ferment.

(*c.*) The presence of a little *free fatty acid* greatly favours emulsification. Take two samples of cod-liver oil, one *perfectly* neutral (by no means easily procured), and an ordinary brown oil—*e.g.*, De Jongh's. The latter contains much free fatty acid. Place 5 cc. of each in two test-tubes, and pour on them a little solution of sodic carbonate (1 per cent.) The neutral oil is not emulsified, while the rancid one is at once, and remains so. Many oils that do not taste rancid contain free fatty acids, and only some of them give up their acid to water, just according as the fatty acid is soluble in water, or not.

7. B. The Fat-Splitting Action of Pancreatic Juice.

(*a.*) **Prepare a perfectly neutral oil.**—A perfectly *neutral* oil is required, and as all commercial oils contain free fatty acids, they must not be used. Place olive or almond oil in a porcelain capsule, mix it with not too much baryta solution, and boil for some time. Allow it to cool. The unsaponified oil is extracted with ether, the ethereal extract separated from the insoluble portion, and the ether evaporated over warm water. The oil should now be perfectly neutral (*Krukenberg*).

(*b.*) Mix the oil with finely-divided, perfectly fresh pancreas (not a watery extract), and keep it at 40° C. After a time its reaction becomes acid, owing to the formation of a fatty acid. This experiment is by no means easy to perform, and some observers deny altogether the existence of a fat-splitting ferment. The free fatty acids thus liberated unite with the alkaline bases of bile, and form soaps.

8. IV.—Milk-Curdling Ferment.

(*a.*) Add a drop or two of the brine extract of the pancreas prepared for you, to 5 cc. of warm milk in a test-tube, and keep it at 40° C. Within a few minutes a solid coagulum forms, and thereafter the whey begins to separate.

(*b.*) Repeat (*a.*), but add a grain or less of bicarbonate of soda to the milk. Coagulation occurs just as before, so that this ferment is active in an alkaline medium.

(*c.*) Boil the ferment first. Its power is instantly destroyed.

9. Action on Milk.

(*a.*) Dilute cow's milk with 5 volumes of water. Test a portion, and note that acetic acid throws down a flocculent precipitate of casein. Place some of the diluted milk in a test-tube, add a drop or two of pancreatic extract, or the Liquor Pancreaticus of Benger. Expose on a water-bath at 40° C. for half an hour. Note that the casein is first curdled and then dissolved, and as this occurs, the milk changes from a white to a yellowish colour.

(*b.*) Divide the fluid of (*a.*) into two portions, A and B. To A add dilute acetic acid, there is no precipitation of casein, which has been converted into peptones. To B add caustic soda and dilute copper sulphate, which give a rose colour, proving the presence of peptones.

10. To Peptonise Milk.—A pint of milk is diluted with a

quarter of a pint of water, and heated to a luke-warm temperature, about 140° F. (or the diluted milk may be divided into two equal portions, one of which may be heated to the boiling point and then added to the cold portion, the mixture will then be of the required temperature). Two tea-spoonfuls of Liquor Pancreaticus, together with about fifteen grains or half a level tea-spoonful of bicarbonate of soda are then mixed therewith. The mixture is next poured into a jug, covered, and placed in a warm situation to keep up the heat. In a few minutes a considerable change will have taken place in the milk, but in most cases it is best to allow the digestive process to go on for ten or twenty minutes. The gradually-increasing bitterness of the digested milk is unobjectionable to many palates ; a few trials will, however, indicate the limit most acceptable to the individual patient ; as soon as this point is reached, the milk should be either used or boiled to prevent further change. From ten minutes to half an hour is the time generally found sufficient. It can then be used like ordinary milk.

LESSON IX.

THE BILE.

1. Use ox bile obtained from the butcher.

(*a.*) Observe the **colour** of bile of man and that of the ox, the former is a brownish-yellow, the latter greenish, but often it is reddish-brown when it stands for a short time.

(*b.*) Dilute bile and test its **reaction** = **alkaline** or neutral.

(*c.*) Pour some ox bile from one vessel to another, and note that it is sticky, strings of **mucin** connecting the one vessel with the other.

(*d.*) Dilute bile with 5 volumes of water, add dilute acetic acid, which precipitates the **mucin** coloured with the pigments. Or, dilute bile with its own volume of water, and precipitate the mucin with alcohol. Filter, and observe that the filtrate is no longer sticky, but flows like a watery non-viscid fluid. The mucin remaining on the filter may be washed with dilute spirit and dissolved in lime water.

(*e.*) Bile gives no reactions for albumin.

(*f.*) Fresh human bile gives no spectrum, although the bile of the ox, mouse, and some other animals does.

(*g.*) Bile, besides the ordinary **salts**, contains so much **iron** as to give the ordinary reactions for iron. To bile add hydrochloric acid and potassic ferrocyanide. A blue colour indicates the presence of iron. It is better to use the filtrate of (*d.*) free from mucin.

If fresh bile is not obtainable, use a watery solution of the " Fel bovinum " of the Pharmacopœia.

2. **To Prepare the Bile Salts or Bilin** (glycocholate and taurocholate of sodium).

(*a.*) Mix ox bile with animal charcoal in a mortar to form a thick paste. Evaporate to complete dryness over a water-bath. Keep this for preparing an alcoholic solution of the bile salts.

(*b.*) Take some of the dry charcoal-bile mixture, and add five times its volume of absolute alcohol. Cork the flask or test-tube in which the mixture is placed. Shake up the mixture from time to time, and after half an hour, filter. To the filtrate add much ether, which gives a white precipitate of the bile salts. If no water be present, sometimes the bile salts are thrown down crystalline ; but not unfrequently they go down merely as a milky opalescence, which quickly forms resinous masses. It is best to allow the mixture to stand for a day or two, to obtain the large glancing needles which constitute **Plattner's Crystallised Bile.**

3. Pettenkofer's Test for Bile Acids and Cholic Acid.

(*a.*) Place some bile in a test-tube, add a drop or two of syrup of cane sugar, and mix. Pour in *concentrated* sulphuric acid, and at the line of junction of the two fluids a purple colour is obtained.

The white deposit seen above the line of junction is precipitated bile acids. They are insoluble in water.

(*b.*) Or, after mixing the syrup with the bile, add the strong sulphuric acid drop by drop, mixing it thoroughly. Heat gently, and the fluid becomes a deep purple colour. Take care not to add too much syrup, and not to overheat the tube. If the requisite amount of sulphuric acid be added, the temperature becomes sufficiently high ($70°$ C.) without requiring to heat the tube.

(*c.*) A better way of doing the test is as follows :—After mixing the bile and syrup, shake the mixture until the upper part of the tube is filled with froth. Pour sulphuric acid down the side, and a purple-red colour is struck in the froth.

(*d.*) It may also be done by mixing a few drops of bile in a porcelain capsule, with a small crystal of cane sugar, and after the sugar is dissolved, adding sulphuric acid drop by drop.

(*e.*) **Strasburg's Modification** (*e.g.*, for bile in urine).—To the urine add a little syrup and mix. Dip filter-paper into the fluid and dry the paper. On placing a drop of sulphuric acid on the latter, after some time, a purple spot which has eaten into the paper is observed.

(*f.*) Repeat any or all of the above processes with a watery solution of the bile salts.

4. Similar colour reactions are obtained with many other substances—*e.g.*, albumin and fats.

Albumin and Sulphuric Acid.—To a solution of syntonin and syrup add strong sulphuric acid, and a similar tint is obtained. The spectra, however, are different, the red-purple fluid from bile gives two absorption bands, one between E and F, and another between D and E. In the albuminous solutions, only one absorption band exists between E and F.

5. Action of Bile or Bile-Salts in Precipitating Sulphur.

(*a.*) Take two beakers and in one (A) place diluted bile, and in the other (B) water. Pour flowers of sulphur on the surface of the water of both. The sulphur falls in a copious shower through the fluid of A, while none passes through B.

(*b.*) Test to what extent bile may be diluted before it loses this property, which is due to the diminution of the surface tension by the bile-salts (*M. Hay*).

(*c.*) Perform the same experiment with a solution of the bile-salts.

Bile Pigments.—The chief are **bilirubin** (red), **biliverdin** (green), and **urobilin.**

6. Gmelin's Test for the Bile Pigments.

(*a.*) Place a few drops of bile on a *white* porcelain slab. With a glass rod, place a drop or two of strong *nitric acid containing nitrous acid* near the drop of bile, bring the acid and bile into contact, when there is immediately a play of colours, beginning with *green* and passing into blue, violet, red, and dirty yellow.

(*b.*) Place a little nitric acid in a test-tube. Slant the tube and pour in bile, a similar play of colours occurs—

green above, blue, violet, red, and yellow below. It is better to do this reaction after removal of the mucin by acetic acid (Lesson IX., 1, *d*). Or add the nitric acid, and shake after the addition of every few drops, the successive colours from green to yellow are obtained in great beauty.

(*c*.) For a modification applicable to urine, see "Urine."

7. Cholesterin and Gall Stones.

(*a*.) **Preparation.**—Powder a gall-stone and extract it with ether. Heat the test-tube with ether in warm water, and see that no gas is burning near it. Allow a drop of the ethereal solution to evaporate on a glass-slide or watch-glass, and examine the crystals under the microscope. They are flat plates, with an oblong piece cut out of one corner (Fig. 11).

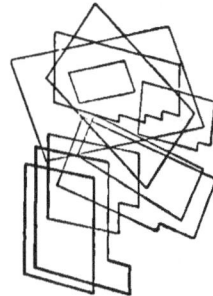

(*b*.) Heat a few crystals in a watch-glass with a few drops of moderately strong sulphuric acid, and then add iodine; a play of colours, passing through violet, blue, green, red, and brown, occurs. (It requires a little practice to get this test always to succeed.)

Fig. 11.—Crystals of Cholesterin.

(*c*.) Dissolve some crystals in chloroform, add an equal volume of *concentrated* sulphuric acid, and shake the mixture. When the chloroform solution floats on the top, it becomes blood-red, but changes quickly on exposure to the air, passing through violet and blue to green and yellow. A trace of water decolourises it at once. The layer of sulphuric acid shows a green fluorescence. [This reaction is not easily performed.]

(*d*.) The crystals when acted on by strong sulphuric acid become red. Do this on a slide under the microscope.

(*e*.) Place a crystal on a piece of porcelain, add a drop of strong nitric acid, evaporate to dryness at a gentle heat, until a yellow residue is obtained. A drop of ammonia

5

strikes a red colour, which is not altered by the addition of caustic soda.

(*f.*) Examine microscopically crystals of cholesterin floating in hydrocele fluid. The crystals may not be quite perfect, but their characters are quite distinct.

8. Action of Bile in Digestion.

(*a.*) **Action on Starch.**—Test if bile converts starch mucilage into a reducing sugar, as directed for saliva (Lesson VI., 3).

(*b.*) **Action on Fats.**—Mix 10 cc. of bile with 2 cc. of almond oil or oleic acid. Shake them together, and observe both by the naked eye and the microscope to what extent emulsion occurs, and how long it lasts. Compare a similar mixture of oil and water. In the former case a pretty fair emulsion will be obtained. In the latter the oil and water separate almost immediately.

(*c.*) Mix 10 cc. of bile with 2 cc. of almond oil, to which some oleic acid is added. Shake well, and keep the tube in a water bath at 40° C. A very good emulsion is obtained. The bile dissolves the fatty acids, and the latter decompose the salts of the bile acids; the bile acids are liberated, while the fatty acid unites with the alkali of the bile-salts to form . a soap. The soap is soluble in the bile, and serves to increase the emulsifying power, as an emulsion once formed lasts much longer in a soapy solution than in water.

(*d.*) **Favours Filtration and Absorption.**—Place two small funnels exactly the same size in a filter-stand, and under each a beaker. Into both funnels put a filter-paper; moisten the one with water (A), and the other with bile (B); pour into both an equal volume of almond oil; cover with a slip of glass to prevent evaporation. Set the whole aside for twelve hours, and note that the oil passes through the filter B, but scarcely any through A.

(*e.*) **Effect on the Proteid Products of Gastric Digestion.**—Digest some fibrin in artificial gastric juice, filter, and to the filtrate add drop by drop some ox bile, or a solution of bile-salts. It causes a white precipitate of peptones and parapeptones. The acid of the gastric juice splits up the bile-salts, so that the bile acids are also thrown down.

(*f.*) **Action on Syntonin.**—Prepare syntonin in solution (Lesson I., 7, *f*), and add a few drops of bile or bile-salts. It causes a curdling of the whole mass. Be careful not to add too much bile. In (*e.*) and (*f.*) it is better to add the bile-salts, because the free hydrochloric acid gives a precipitate with bile.

LESSON X.

GLYCOGEN IN THE LIVER.

1. Preparation.

(*a.*) Boil some water slightly acidulated with acetic acid, and keep it boiling. Feed a rat, and three or four hours thereafter decapitate it. *Rapidly* open the abdomen, remove the liver, cut one half of it in pieces, and throw it into the boiling acidulated water. Lay the other half aside, keeping it moist in a warm place for some hours. After boiling the first portion for a time, pound it in a mortar with sand, and boil again. Filter while hot. The filtrate is milky or *opalescent*, and is a watery solution of glycogen. The acetic acid coagulates the proteids, while the boiling water destroys a ferment in the liver, which would convert the glycogen into grape-sugar.

(*b.*) Feed a rabbit on carrots, and after from two to three hours decapitate it. Open the abdomen, tear out the liver, cut it rapidly in pieces, and take one half—laying the other half aside as in (*a.*)—throw it into boiling water, boil it, and afterwards pound it in a mortar and boil again. Filter while hot, and observe the *opalescent* filtrate, which is a solution of glycogen and proteids. The filtrate should flow into a cooled beaker, placed in a mixture of ice and salt. Precipitate the proteids by adding alternately hydrochloric acid and potassio-mercuric iodide, until all the proteids are precipitated. Filter off the proteids, and the beautiful opalescent filtrate is an imperfect solution of glycogen (**Brücke's method**).

(*c.*) Instead of a rat or rabbit's liver, use oysters, and prepare a solution of glycogen by methods (*a.*) or (*b.*)

(*d.*) Use the other half of the liver of the rat or rabbit that has been kept warm, and make a similar extract of it.

2. Precipitate the Glycogen.—Evaporate the filtrate of (*a.*) or (*b.*) to a small bulk, and precipitate the glycogen as a white powder by adding a large amount of alcohol.

3. Preparation of Potassio-mercuric Iodide. — Precipitate a saturated solution of potassic iodide with a similar solution of mercuric chloride ; wash the precipitate, and dissolve it to saturation in a hot solution of potassic iodide.

4. Tests.

(*a.*) To the opalescent filtrate add a solution of iodine= a port-wine red colour (like that produced by dextrin). If much glycogen be present the colour disappears, and more iodine has to be added. Heat the red fluid ; the colour disappears on heating, but reappears on cooling.

N.B.—In performing this test, make a *control-experiment.* Take two test-tubes, A and B. To A add solution of glycogen ; to B, an equal volume of water. To both add the same amount of iodine solution. A becomes red, while B is but faint yellow.

(*b.*) Test a portion of the glycogen solution for grape-sugar. There should be none, or only the faintest trace.

(*c.*) To a portion of the glycogen solution add saliva or Liquor Pancreaticus (A), and to another portion add blood (B), and place both on a water-bath at 40° C. After ten minutes test both for sugar. (A) will now be transparent, and give no reaction with iodine. Perhaps both will give the reaction; but certainly (A) will, if care be taken that the solution is not acid, after adding the saliva. The ptyalin converts the glycogen into a reducing sugar.

(*d.*) Boil some glycogen solution with dilute hydrochloric

acid in a flask ; neutralise with caustic soda, and test with Fehling's solution for sugar.

(*e.*) To another portion add plumbic acetate = a precipitate (unlike dextrin).

(*f.*) To another portion add plumbic acetate and ammonia = a precipitate (like dextrin).

5. Test the watery extract of the second half of the liver (*d*).

(*a.*) It will perhaps give no glycogen reaction, or only a slight one.

(*b.*) It contains much reducing sugar.

6. Extract of a Dead Liver.

(*a.*) Mince a piece of liver from an animal (*e.g.*, an ox) which has been dead for twenty-four hours. Place the finely-divided liver in water or in a saturated solution of sodic sulphate, and boil to make a watery extract. Filter, and observe the filtrate ; it is clear and yellowish in tint, but not opalescent.

(*b.*) Test its reaction = acid.

(*c.*) Test with iodine after neutralisation with sodic carbonate and filtration = no glycogen.

(*d.*) Test for grape-sugar = much sugar.

After death the glycogen is rapidly transformed into grape-sugar.

LESSON XI.

MILK, FLOUR, AND BREAD.

1. **Milk**—Use fresh cow's milk.

(*a.*) Examine a drop of milk under the **microscope**, noting the numerous small, highly-refractive **fat globules** floating in a fluid (Fig. 12, M).

(i.) Add a drop of acetic acid, and observe how the globules run into groups.

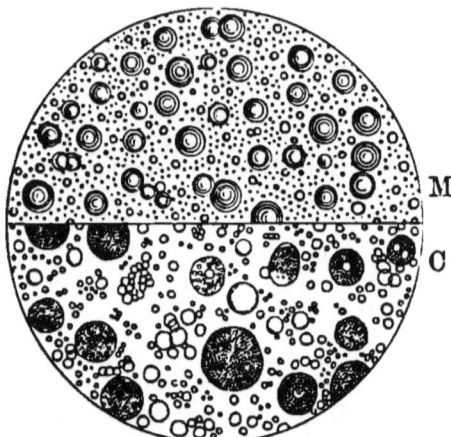

(ii.) To a fresh drop add osmic acid, and observe how the globules first become brown and then black.

(iii.) If a drop of colostrum is obtainable, examine for the "colostrum corpuscles" (Fig. 12, C).

(b.) Examine the "naked-eye" characters of milk.

Fig. 12.—Microscopic appearance of milk.— The upper half, M, is milk; the lower half colostrum, C.

(c.) Test its reaction with litmus-paper. It is usually alkaline.

(d.) Take the **specific gravity** of perfectly fresh unskimmed milk with a hydrometer or lactometer. It is usually between 1025 – 1030. Take the specific gravity next day after the cream has risen to the surface. The specific gravity is higher.

(e.) Dilute milk with ten times its volume of water, and carefully neutralise it with *dilute* acetic acid, observe that there is no precipitate, as the casein is prevented from being precipitated by the presence of alkaline phosphates (Lesson I., 7). Cautiously add the acetic acid until there is a copious granular-looking precipitate of **casein**, which as it falls, entangles the greater part of the fat in it. If desired the precipitation is hastened by heating the fluid to 70° C.

(f.) Filter (e.) through a moist-plaited filter. Keep the residue on the filter. The filtrate of (f.) should be clear. Divide it into two portions. Take one of the portions, divide it into two and boil one = a precipitate of **serum-albumin.** Filter, and keep the filtrate to test for sugar. To the remainder add potassic ferrocyanide, which also precipitates serum-albumin.

(*g.*) With the second half of the filtrate test for milk-sugar or **lactose** with Fehling's solution, or by Trommer's test (Lesson II., 6; IV., *b.*, *c.*) Instead of proceeding thus, test for the presence of a reducing sugar with the filtrate of (*f.*) after the separation of the serum-albumin.

(*h.*) Scrape off the residue of casein and fat from the filter (*f.*) ; wash it with water from a wash bottle, and exhaust the residue with a mixture of ether and alcohol. On placing some of the ethereal solution on a slide, and allowing it to evaporate, a greasy stain of **fat** is obtained.

(*i.*) To *fresh* milk add a drop of tincture of guaiacum, which strikes a blue colour ; boiled milk does not do so.

2. Separation of the Casein by Salt.—To one volume of milk in a test-tube, add two volumes of a saturated solution of common salt, and then excess of powdered salt, (or magnesic sulphate may be used). Shake the tube vigorously for a time, when the casein and fat separate out, rise to the surface, and leave a clear fluid or whey behind. This fluid contains the lactose, salts, and serum-albumin.

3. Separation of the Casein and Fat by Filtration.—Using a Bunsen's pump, filter milk through a porous cell of porcelain. The particulate matters—casein and fat— remain behind, while a clear filtrate containing the other substances in milk passes through. The porous cell is left empty and fitted with a caoutchouc cork with two glass tubes tightly fitted into it. One tube is closed with a clip (Fig. 13), and the other is attached to the pump. Place the porous cell in an outer vessel containing milk. On exhausting the porous cell, a clear watery fluid slowly passes through. Test it for proteids and sugar. Notice the absence of fat and casein.

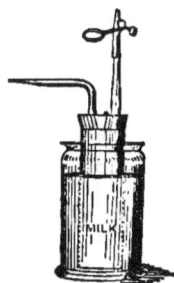

Fig. 13. — Porous Cell for the Filtration of Milk.

4. Souring of Milk.—Place a small quantity of milk in a vessel in a warm place for several days, when it turns sour and curdles. It becomes acid—test this [Lesson VII., 8, (e.)]—having undergone the **lactic acid fermentation,** the lactose being split up by a micro-organism into lactic acid.

5. To Separate the Butter.—Place a little milk in a narrow, cylindrical, stoppered bottle; add half its volume of caustic soda and some ether, and shake the mixture. Put the bottle in a water-bath at a low temperature; the milk loses its white colour, and an ethereal solution of the fats floats on the surface. On evaporating the ethereal solution, the butter is left behind.

6. Curdling of Milk.

(*a.*) **By an Acid.**—Place some milk in a flask; warm it to 40° C., and add a few drops of acetic acid. The mass clots or curdles, and separates into a solid **curd** (casein and fat), and a clear fluid, the **whey,** which contains the lactose. Filter.

(*b.*) **By Rennet-Ferment.**—Take 5 cc. of fresh milk in a test-tube, heat it in a water-bath to 40° C., and add to it a small quantity of extract of rennet, or an equal volume of a glycerin extract of the gastric mucous membrane, which has been neutralised with dilute sodic carbonate, and place the tube again in the water-bath at 40° C.

Observe that the whole mass curdles in a few minutes, so that the tube can be inverted without the curd falling out. By-and-by the curd shrinks, and squeezes out a clear slightly-yellowish fluid, the whey. Filter.

(*c.*) Using commercial rennet-extract, repeat (*b.*), but boil the rennet first, it no longer effects the change described above. The rennet-ferment is destroyed by heat.

(*d.*) Boil the milk and allow it to cool; then add rennet; in all probability, no coagulation will take place. Boiled milk is far more difficult to coagulate with rennet than unboiled milk.

(*e.*) Take some of the curd of 6 (*a.*) Dissolve one **part** in caustic soda and the other in lime-water. Add rennet to both, warm to 40° C. The lime solution coagulates, the soda solution does not.

7. The ordinary Salts are present.

(*a.*) Using the filtrate of 6 (*a.*) add magnesia mixture—Lesson XIV., 6 (*g.*)—*i.e.*, ammonio-sulphate of magnesia, which gives a precipitate of *phosphates.*

(*b.*) Silver nitrate gives a precipitate insoluble in nitric acid, indicating *chlorides.*

8. Boil milk in a porcelain capsule for a time to cause evaporation. It is not coagulated, but a pellicle of insoluble casein forms on the surface. Remove it and boil again; another pellicle is formed.

9. Opacity of Milk. Vogel's Lactoscope.

Apparatus required.—A graduated cylindrical cc. measure to hold 200 cc.; a lactoscope, with parallel glass sides, 5 mm. apart (Fig. 14); a burette or pipette, finely graduated; a stearin candle.

Method.—(*a.*) Be certain, by microscopical examination, that the milk contains no starch, or chalk, or other granular impurity.

Fig. 14.—Lactoscope.

(*b.*) Place 100 cc. of water in the cylindrical measuring glass, and add 3 cc. of milk. Mix thoroughly.

(*c.*) In a dark room place the lactoscope on a table, and 1 metre distant from it a lighted stearin candle. Fill the lactoscope with the diluted milk, and look at the candle flame through the glass. If the contour of the flame can be seen distinctly, pour back the diluted milk into the bottle, and add another 3 cc. of milk. Mix again. Test the mixture again, and repeat until, on looking through the glass, the outline of the candle flame can no longer be recognised. Add together the quantities of milk used. An empirical table constructed by Vogel, gives the percentage of fat.

10. Wheaten Flour.

(*a.*) **Gluten.**—Moisten some flour with water until it forms a tough tenacious dough; tie it in a piece of muslin, and knead it in a vessel containing water until all the starch is separated. There remains on the muslin a greyish-white, sticky elastic mass of "**crude gluten,**" consisting of the insoluble albumenoids, some of the ash, and the fats. Draw out some of the gluten into threads, and observe its tenacious characters.

(*b.*) Dry some of the gluten, and heat it strongly in a test-tube; an ammoniacal odour similar to that of burned feathers is evolved. Water, which is alkaline (due to ammonia), condenses in the upper part of the tube.

(*c.*) Extract 10 grms. of wheaten flour with 50 cc. of water in a large flask. Shake it from time to time, and allow it to stand for several hours. Filter. If the filtrate is not quite clear, filter again. Heat a part of the clear filtrate, and observe the coagulation of *vegetable albumin.*

(*d.*) Test another portion of the filtrate from (*c.*) for the xantho-proteic reaction.

(*e.*) Another portion of (*c.*) is to be precipitated by acetic acid and ferrocyanide of potassium.

(*f.*) Test a third portion of (*c.*) for the biuret reaction. This is best seen on slightly heating. Take care not to boil the liquid, or the reaction for sugar will be got instead.

(*g.*) Extract some wheaten flour with a 10 per cent. solution of common salt for twelve hours. Filter, and drop some of the clear filtrate into a large vessel of water; a milky precipitate of a *globulin* is obtained.

(*h.*) On saturating some of the filtered saline extract of (*g.*) with powdered NaCl or $MgSO_4$, a precipitate of a globulin is thrown down.

(*i.*) **Fats.**—Shake up some wheaten flour with ether in a cylindrical stoppered vessel or test-tube, with a tight-fitting cork. Allow the mixture to stand for an hour, shaking it from time to time. Filter off the ether; place some of it on a perfectly clean watch-glass, and allow it to evaporate spontaneously, when a greasy stain will be left.

(*j.*) The chief mineral matter, or **salts**, consists of potassium and phosphoric acid. The watery extract gives a yellow precipitate with platinic chloride, showing the presence of potassium; while heating it with molybdate of ammonium and nitric acid gives a canary-yellow precipitate, proving the presence of phosphates.

11. Pea Meal.

(*a.*) Make corresponding watery and saline extracts, and perform the same experiments with them as in Lesson XI., 10, (*c.*), (*d.*), (*e.*), (*f.*), (*g.*), (*h.*)

(*b.*) Observe the copious precipitate on boiling the watery extract.

(*c.*) Note specially the copious deposit of globulin on adding the saline extract to water.

12. Bread.

(*a.*) Make a watery extract with warm water, filter, and test the filtrate. Its reaction is alkaline.

(*b.*) Test for starch and sugar.

LESSON XII.

MUSCLE.

1. The Reaction.

(*a.*) Arrange two strips of glazed litmus-paper, one red and one blue, side by side. Pith a frog; cut out the gastrocnemius, remove as much blood as possible, divide the muscle transversely, and press the cut ends on the litmus-paper; a faint blue patch is produced on the red paper, showing that the muscle is *alkaline* during life. The blue paper is not affected.

(*b.*) Have some water at 50° C. Dip into it the other gastrocnemius until *rigor mortis* sets in. Test its reaction; now it is *acid*.

(*c.*) Boil some water, and plunge into it any other muscle of the same frog ; it is *alkaline.*

(*d.*) Test the reaction of a piece of butcher's-meat. It is intensely acid.

(*e.*) Tetanise a muscle for a long time. Its reaction becomes acid.

2. Watery and Saline Extracts.

(*a.*) Mince some perfectly fresh muscles from a rabbit or dog. Extract with water, stirring from time to time. After half an hour, pour off, and filter the watery extract. Re-extract the remainder with water until the extract gives no proteid reactions. For the purposes of this exercise, half an hour is sufficient. Keep the filtrate, which contains the substances *soluble in water.*

(*b.*) Take some perfectly fresh muscle from a rabbit, rub it up with sand in a mortar, and extract it with a large volume of 10 per cent. solution of ammonium chloride, NaCl, or 5 per cent. $MgSO_4$. Stir occasionally, and allow it to extract for an hour. A stronger extract is obtained if it be left until next day. Pour off the fluid, keep it as it contains the substances *soluble in saline* solutions, the **globulins.**

3. With the filtrate of 2 (*a.*)

(*a.*) Test for proteids—*e.g.*, serum-albumin.

(*b.*) Test the coagulating point of the proteids it contains (45° and 75° C.)

(*c.*) Add crystals of ammonium sulphate to saturation, which precipitates all the proteids.

4. With the filtrate of 2 (*b.*)

(*a.*) Pour a few drops into a large quantity of water, observe the milky deposit of **myosin.** The precipitate is redissolved by adding a strong solution of common salt.

(*b.*) Test the coagulating point. Four proteids are coagulated by heat at 47°, 56°, 63°, and 73° C., an albumose being left in solution. The fluid is acid in reaction.

(*c.*) Saturate the filtrate with crystals of sodic chloride or ammonium chloride. The myosin is precipitated.

(*d.*) Collect some of the precipitate of 4 (*c.*), dissolve it with a weak solution of common salt, and test for proteid reactions (Lesson I., 1). Repeat 3 (*c.*)

5. The Extractives of Muscle. Prepare **Creatin**, omitting the others.

(*a.*) Make a strong watery solution of Liebig's extract of meat. Cautiously add lead acetate until precipitation ceases, avoiding excess of the lead. Filter, pass sulphuretted hydrogen through the filtrate to get rid of the lead. A pellicle is very apt to form on the surface. Filter, and evaporate the filtrate to a syrup on a water-bath, and set it aside in a cool place to crystallise. Crystals of creatin separate out.

(*b.*) After several days, when the creatin has separated, pour off the mother liquor, add to it 5 volumes of 90 per cent. alcohol to precipitate more creatin. Filter, wash the crystals with alcohol, redissolve them in a boiling water, allow them to recrystallise, and examine them with the microscope.

Sarkin and xanthin may be prepared from the alcoholic filtrate of (*b.*)

LESSON XIII.

THE URINE.

1. The Urine is a transparent light straw or amber-coloured watery secretion derived from the kidneys, containing nitrogenous matters, salts, and gases: it has a peculiar odour, bitter saltish taste, and acid reaction.

(*a.*) Evaporate a drop of urine on platinum foil. Do this over the flame of a Bunsen burner, taking care not to burn it. A brownish-yellow stain, with an ammoniacal odour, is obtained.

(*b.*) Evaporate a drop of distilled water to compare with this—no residue.

(*c.*) Burn the stain of 1 (*a.*) in the flame: it chars, indicating the presence of *organic* matter, while a faint trace of ash or *inorganic* matter is left.

2. **The Quantity.**—*Normal.*—About 2½ pints (50 ounces), or 1500 cc., in twenty-four hours, although there may be a considerable variation even in health, the quantity being regulated by the amount of fluid taken, and controlled by the state of the tissues, the pulmonary and cutaneous excretions.

Increased by drinking water (*Urina potus*) or diuretics; when the skin is cool, its blood-vessels are contracted, and the cutaneous secretion is less active; after a paroxysm of hysteria, and some convulsive nervous diseases; in *Diabetes insipidus* and *D. mellitus;* some cases of hypertrophy of the left ventricle, and some kidney diseases. The increase may be temporary or persistent, the former as the effect of cold, diuretics, or nervous excitement; the latter in diabetes and certain forms of kidney disease.

Diminished after profuse sweating, diarrhœa; early stage of acute Bright's disease; some forms of Bright's disease; the last stages of all forms of Bright's disease; in general dropsies; in acute febrile and inflammatory diseases.

3. **The Colour.**—*Normal.*—Light straw to amber-coloured. The colour varies greatly even in health, and is due to the presence of pigments, probably largely derived from the decomposition of hæmoglobin. The colour largely depends on the degree of dilution of the urine pigments.

Pale after copious drinking, in diabetes, anæmia, and chlorosis; after paroxysmal nervous attacks (hysteria). *N.B.*—Pale urines indicate the absence of fever.

High-coloured after severe sweating, violent muscular exercise, diarrhœa, or during febrile conditions.

Pathological pigments, purpurine or uro-erythrine in febrile disorders; bile pigments; blood.

Medicinal Substances.—Creasote and carbolic acid make urine nearly black. This is due not to carbolic acid, but to hydrochinon. Sometimes these urines become almost black on standing exposed to the air. Rhubarb (gamboge-yellow); senna (brownish).

4. **The Specific Gravity.**—*Normal.*—Specific gravity 1020 (1018-1025).

To take the Specific Gravity.—This is usually done by means of the **urinometer** (Fig. 15). The instrument ought to be tested by placing it in a cylindrical vessel filled with distilled water to ascertain that its zero is correct.

> (*a.*) Fill a tall cylindrical vessel with urine, and place the urinometer in it. Bring the vessel to the level of the eye, and as soon as the instrument comes to rest, read off the mark on its stem opposite the lower surface of the meniscus against a bright background.

Precautions.—1. The vessel must be so wide that the urinometer can float freely and not touch the sides. 2. The instrument must be dry before being placed in the fluid. 3. The urine itself must be clear, and free from air-bubbles on the surface; the latter can be readily removed by means of a fold of blotting-paper. *N.B.*—It is always necessary to take the specific gravity of the "mixed" urine of twenty-four hours.

Low Specific Gravity.—All causes which increase the water of the urine only—*e.g.*, drinking on an empty stomach; after hysteria; in *Diabetes insipidus* or *Polydipsia*. *N.B.*—If continually below 1015, suspect *Diabetes insipidus* or chronic Bright's disease.

High Specific Gravity.—When the urine is concentrated; *Diabetes mellitus*, due to a large amount of grape sugar; first stages of acute fevers; rapid wasting of the tissues, especially if associated with sweating or diarrhœa. It is highest normally three to four hours after a meal; and as it varies during the day, it is necessary to mix the urine of the twenty-four hours, and test the specific gravity of a sample of the "mixed urine." *N.B.*—If above 1025 and the urine be pale, suspect saccharine diabetes.

Fig. 15.
Urinometer.

5. Determination of the amount of Solids from the Specific Gravity.—By *Christison's formula* ("*Häser-Trapp's coefficient*") "multiply the last two figures of a specific gravity expressed in four figures by 2·33. This gives the quantity of solid matter in every 1000 parts"—*i.e.*, the number of grammes in 1000 cc. (33⅓ oz.)

Example.—Suppose a patient to pass 1200 cc. of urine in twenty-four hours, and the specific gravity to be 1022, then

$$22 \times 2\cdot33 = 51\cdot26 \text{ grms. in } 1000 \text{ cc.}$$

To ascertain the amount in 1200 cc.

$$1000 : 1200 :: 51\cdot26 : x = \frac{51\cdot26 \times 1200}{1000} = 61\cdot51 \text{ grms.}$$

This formula is purely empirical, and is not applicable where the variations are very marked, as in saccharine diabetes and some cases of Bright's disease, where there is a great diminution of urea.

The normal quantity of solids, or the **total solids**—sometimes spoken of as "solid urine"—is about 70 grms. in twenty-four hours—*i.e.*, 1000 to 1050 grains. Parkes gives an average of 945 grains per day for an average adult male between twenty and forty years of age. The latter estimate gives about 20 grains of solids per fluid ounce of urine, or about 4 per cent. of solids.

6. The Odour is "peculiar" and "characteristic," somewhat aromatic in health. Certain medicinal and other substances influence it—turpentine (violets); cubebs, copaiba, and sandal wood oil give a characteristic odour, and so do asparagus, valerian, assafœtida, garlic, &c. In *disease*, note the ammoniacal odour of putrid urine and the so-called "sweet" odour in saccharine diabetes.

7. The Reaction.—*Normal.*—Slightly acid, it turns blue litmus-paper slightly red, and does not affect red litmus-paper. The acidity is chiefly due to acid sodium phosphate (NaH_2PO_4), acid urates, and very slightly to free acids—lactic, acetic, oxalic, &c. A *neutral* urine does not alter either blue or red litmus paper. A *very acid* urine turns blue litmus paper very red. Sometimes *violet* litmus paper is used; it becomes red in acid urine and blue in alkaline.

(*a.*) Take two pieces of neutral litmus-paper, put a drop of water on the one, and a drop of urine on the other, and observe the effects.

(*b.*) Test with appropriate litmus-paper a normal, very acid, neutral, and alkaline urine.

(*c.*) Test also with violet litmus-paper.

8. Variations during the Day.—Two or three hours after a meal the urine becomes neutral or alkaline. (See that the bladder be emptied before beginning the experiment.) The cause of the alkalinity is a fixed alkali, probably derived from the basic alkaline phosphates taken with the food (Roberts).

Nature of the Food.—With a vegetable diet, the excess of alkali causes an alkaline urine. In herbivora it is alkaline, in carnivora very acid. Herbivora (rabbits) whilst fasting have a clear acid urine, because they are practically living on their own tissues. Perhaps this is one of the reasons why the urine is so acid in fevers. Inanition renders the urine *very acid* (Chossat).

Medicines.—*Acids* slightly increase the acidity. *Alkalies* and their carbonates are more powerful than acids, and soon cause alkalinity; alkalies, *e.g.*, the alkaline salts of citric, tartaric, malic, acetic, and lactic acids, appear as carbonates (Wöhler).

9. The Alkalinity may be due to the presence of a Fixed or a Volatile Alkali.—In the former case, the blue colour of the litmus-paper does not disappear on heating; in the latter it does, and the paper assumes its original red colour.

(*a.*) Test with two pieces of red litmus-paper two samples of urine, one alkaline from a fixed alkali, and the other from a volatile one. Both papers become blue.

(*b.*) Place both side by side on a glass slide, heat them carefully, and note that the blue colour of the one disappears (volatile alkali), the red being restored, while the blue of the other remains (fixed alkali).

The **alkalinity** may be caused by the presence of ammonium carbonate (volatile), derived from the decomposition of urea; the urine may be ammoniacal when passed, in which case there is always disease of the urinary mucous membrane; or it may become so on standing—from putrefaction—when it is always turbid, and contains a sediment consisting of amorphous phosphate of lime and triple-phosphate, and sometimes urate of ammonium; it has an offensive ammoniacal odour, and is very irritating to the mucous membrane.

6

The **acidity** is increased during the resolution of febrile diseases; is excessive in gout and acute rheumatism, and whenever much uric acid is given off (uric acid diathesis); in saccharine diabetes, when certain acids are taken with the food (CO_2, benzoic).

The amount of the acidity may be determined by using a standard solution of caustic soda.

10. Fermentation of the Urine.—When urine is freely exposed to the air it undergoes two fermentations—(1) the *acid;* (2) the *alkaline.* The urine at first becomes slightly *more acid,* from the formation of lactic and acetic acids (although this is denied by some observers), then it gradually becomes *neutral,* and finally *alkaline* from putrefaction. It becomes lighter in colour, turbid, and a whitish heavy precipitate occurs; a pellicle forms on the surface, it swarms with bacteria, and it has an ammoniacal odour, which is due to the splitting up of the urea, thus—

$$CON_2H_4 + 2H_2O = (NH_4)_2 CO_3$$

The carbonate of ammonia makes the urine alkaline, and the earthy phosphates are precipitated because they are insoluble in an alkaline urine. The phosphate of lime is precipitated as such, while the phosphate of magnesia unites with the ammonia and is precipitated as ammonio-magnesic phosphate or triple phosphate (Mg $NH_4 PO_4 + 6H_2O$).

It is not known what ferment causes this reaction —whether mucus, bacteria-like organisms, or some amorphous ferment.

Fig. 16. — Deposit in "acid fermentation" of urine.—*a*, Fungus; *b*, amorphous sodium urate; *c*, uric acid; *d*, calcium oxalate.

N.B.—Although urine may be kept "sweet" for a long time in perfectly clean

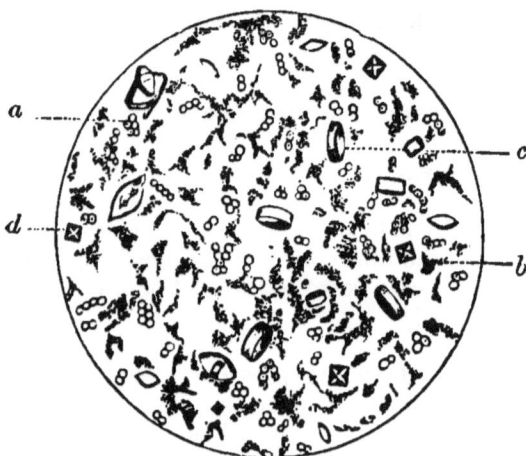

vessels, still when mixed with decomposing matters it has a marked tendency to putrefy.   Insist that all urinary vessels be scrupulously clean ; and that all instruments introduced into the bladder be properly purified by carbolic acid or other germicide.

(*a.*) Place some normal urine aside for some days, preferably in a warm place. Observe it from day to day, noting its reaction, change of colour, transparency, odour, and any deposit that may form in it.   Examine the deposit microscopically (Figs. 16, 17).

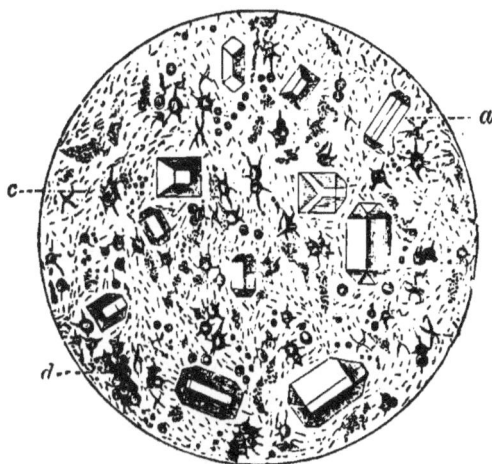

Fig. 17.—Deposit in ammoniacal urine (alkaline fermentation).—*a*, Ammonio-magnesium phosphate; *d*, acid ammonium urate; *c*, bacterium ureæ.

Fermentation is *hastened* by a high temperature, and especially if the urine be passed into a contaminated vessel, or the urine itself contain blood, much mucus or pus.   It is *retarded* in a very acid and concentrated urine.

LESSON XIV.

THE INORGANIC CONSTITUENTS OF URINE.

The ratio of inorganic to organic constituents is 1 to 1·2 − 1·7.

1. **Water** is derived from the food and drink (normal quantity 1500 cc., or about 50 oz.)

2. Chlorides are chiefly those of sodium with a little potassium and ammonium, derived chiefly from the food, and amount to 10 to 15 grammes (150 to 230 grs.)

(*a.*) Test with a few drops of $AgNO_3$ (1 pt. to 8 distilled water) = white, cheesy, or curdy precipitate in lumps insoluble in HNO_3. The phosphate of silver is also thrown down, but it is soluble in HNO_3.

Variations, increased in amount when the urine is secreted in excess, although the NaCl usually remains very constant ($\frac{3}{4}$ per cent.); lessened in febrile affections, and where a large amount of exudation has taken place, as in acute pneumonia, when chlorides may be absent from the urine. The reappearance of chlorides in the urine is a good symptom, and indicates an improvement in the condition of the lung. *N.B.*—The urine ought to be tested daily for chlorides in cases of pneumonia.

(*b.*) Test urine from a case of pneumonia, and compare the amount of the precipitate with that of a normal urine.

Estimation.—A rough estimate may be formed of the amount by allowing the precipitate to subside, and comparing its bulk from day to day.

3. Sulphates are those of soda and potash. Quantity 3 to 4 grammes (46 to 61 grs.) They have no clinical significance.

(*a.*) Test with a soluble salt of barium (the nitrate or chloride) = white heavy precipitate of barium sulphate, insoluble in HNO_3.

4. The Phosphates consist partly of *alkaline* and partly of *earthy* salts in the proportion of 2 to 1. The latter are insoluble in an alkaline medium, and are precipitated when the urine becomes alkaline. They are insoluble in water, but soluble in acids; in urine they are held in solution by free CO_2. The *alkaline* phosphates are very soluble in water, and they *never form urinary deposits.*

5. The Earthy Phosphates are phosphates of lime and magnesia. Quantity 1 to 1·5 grammes (15 to 23 grs.) They are precipitated when the urine is alkaline, although not in the form in which they occur in the urine (Lesson XIII., 10). They are insoluble in water, readily soluble in acetic and carbonic acid, and are precipitated by ammonia.

(a.) To clear filtered urine add nitric acid, boil, and add baric chloride, and boil again = a precipitate of baric sulphate. Filter, and to the cool filtrate add ammonia = a precipitate of baric phosphate.

Clinical Significance.—They are *increased* in osteomalacia and rickets, in chronic rheumatoid arthritis, after prolonged mental fatigue, and by food and drink, and *diminished* in renal diseases and phthisis.

6. The Alkaline Phosphates are chiefly acid sodium phosphate, with perhaps traces of potassium phosphate; they are soluble in water and not precipitated by alkalies, and never occur as urinary deposits. The quantity is 2 to 4 grammes (30 to 60 grs.) They are chiefly derived from the food, and perhaps a small amount from the oxidation of the phosphorus of nerve tissues.

(a.) To fresh, clear filtered urine, add ammonia, caustic soda, or potash, and heat gently until the phosphates begin to separate; let it stand for some time = a white precipitate of the earthy phosphates. Allow it to stand, and estimate approximately the proportion of the deposit. [If a high-coloured urine be used, the phosphates may go down coloured.]

(b.) To urine add about half its volume of nitric acid, and then add solution of ammonium molybdate and boil = a canary-yellow precipitate of ammonium phospho-molybdate.

N.B.—The molybdate is apt to decompose on keeping.

(c.) To urine add half its volume of ammonia, and allow it to stand = a white precipitate of *earthy phosphates.* Filter and test the filtrate as in 6 (b.)

(d.) It gives the reaction for phosphates. This method separates the alkaline from the earthy phosphates.

(e.) To urine add half its volume of baryta mixture [Lesson XVI., 2 (c.)] = a copious white precipitate. Filter and test the filtrate as in 6 (c.) It gives no reaction for phosphoric acid, showing that all the phosphates are precipitated.

(*f.*) To urine add excess of ammonium chloride, and ammonia = a white precipitate of *earthy phosphates* and oxalate of lime. Filter, and to the filtrate add a solution of magnesic sulphate = a precipitate of the alkaline phosphates as *triple phosphate.* If the filtrate be tested for phosphoric acid by **6** (*c.*) no precipitate will be found.

(*g.*) Instead of **6** (*f.*) use **magnesia mixture,** composed of magnesic sulphate and ammonium chloride, each 1 part, distilled water 8 parts, and liquor ammoniæ 1 part. It gives the same result as in **6** (*f.*)

(*h.*) To urine add a few drops of acetic acid, and then uranium acetate or nitrate = bright yellow or lemon-coloured precipitate of uranium and ammonium double phosphate. $2(Ur_2O_3)NH_4PO_4$.

This reaction forms the basis of the process for the volumetric estimation of the phosphoric acid.

7. The other fact connected with the volumetric estimation of phosphoric acid is, that when a uranic salt is added to a solution of potassium ferrocyanide, a reddish-brown colour is obtained.

(**a.**) To a very dilute solution of uranium acetate add potassium ferrocyanide = a brown colour.

8. In some pathological urines the phosphates are deposited on boiling.

(*a.*) Boil such a urine = a precipitate. It may be phosphates or albumin. An albuminous precipitate falls before the boiling point is reached, and phosphates when the fluid is boiled. Add a drop or two of nitric or acetic acid. If it is phosphates, the precipitate is dissolved; if albumin, it is unchanged.

9. Microscopic Examination.—As the alkaline phosphates are all freely soluble in water, they do not occur as a urinary deposit. The earthy phosphates, however, may be deposited.

(*a.*) Examine a preparation or a deposit of **calcic phosphate,** which may exist either in the *amorphous* form, or the crystalline condition, when it is known as *"stellar phosphate."*

(*b.*) **Prepare** "stellar phosphate" crystals by adding some calcium chloride to normal urine, and then nearly neutralising. On standing, crystals exactly like the rare clinical form of stellar phosphate are obtained.

(*c.*) **Triple phosphate** or ammonio-magnesic phosphate never occurs in normal urine, and when it does occur, indicates the decomposition of urea to give the ammonia necessary to combine with magnesic phosphate to form this compound. It forms large, clear "knife-rest" crystals (Fig. 18.)

(*d.*) If ammonia be added to urine, the ammonio-magnesic phosphate is thrown down in a *feathery* form, which is very rarely met with in the investigation of human urine clinically.

10. General Rules for all Volumetric Processes.

(*a.*) The burette must be carefully washed out with the titrating solution, and must be fixed vertically in a suitable holder.

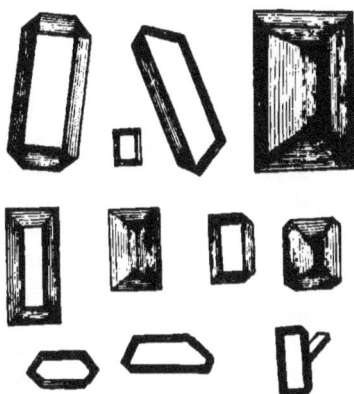

Fig. 18.—Various forms of triple phosphate.

(*b.*) All air-bubbles must be removed from the burette as well as from the outflow tube. The latter must be quite filled with the titrating solution.

(*c.*) Fill the burette with the solution up to zero, and always remove the funnel with which it is filled.

(*d.*) Read off the burette always in the same manner, and always allow a short time to elapse before doing so, in order to allow the fluid to run down the sides of the tube.

(*e.*) The titrating fluid and the fluid being titrated must always be thoroughly well mixed.

11. Volumetric Process for Phosphoric acid, with Ferrocyanide of Potassium as the Indicator.

1 cc. of the SS. (Uranium acetate) = ·005 gramme of phosphoric acid.

(*a.*) Collect and carefully measure the urine passed during 24 hours.

(*b.*) Place 50 cc. of the mixed and filtered urine in a beaker. Do this with a pipette. Place the beaker under a burette.

(*c.*) To the urine add 5 cc. of the solution of sodium acetate; mix thoroughly.

(*d.*) Fill a Mohr's burette with the SS. of uranium acetate up to zero, or to any mark on the burette. See that the Mohr's clip is tight, and that the outflow tube is filled with the SS. Note the height of the fluid in the burette. Heat the urine solution in the beaker to about 80° C. Drop in the SS. ("Standard Solution") of uranium acetate from the burette. Mix thoroughly. Test a drop of the mixture from time to time, until a drop gives a faint brown colour when mixed with a drop of potassium ferrocyanide. Do this on a white plate.

(*e.*) Boil the mixture, and test again. If necessary, add a few more drops of the SS., until the brown colour reappears on testing with the indicator. [Paper may be dipped in the indicator solution and tested with a drop of the mixture.] Read off the number of cc. used.

Example.—Suppose 17 cc. of the SS. are required to precipitate the phosphates in 50 cc. of urine; as 1 cc. of SS. = ·005 gramme of phosphoric acid, then ·005 × 17 = ·085 gramme of phosphoric acid in 50 cc. of urine. Suppose the patient passed 1250 cc. of urine in 24 hours, then $50 : 1250 :: ·085 : x \dfrac{1250 \times ·085}{50} = 2·12$ grammes of phosphoric acid in 24 hours.

12. Solutions Required.

Sodium Acetate Solution.—Dissolve 100 grammes of sodium

acetate in 100 cc. pure acetic acid, and dilute the mixture with distilled water to 1000 cc.

Potassium Ferrocyanide Solution.—Dissolve 1 part of the salt in 20 parts of water.

Uranium Acetate Solution (1 cc. = ·005 gramme H_3PO_4).— Dissolve 20·3 grammes of pure uranium oxide in strong acetic acid; dilute with distilled water to a litre. The strength of this solution must be ascertained by means of a standard solution of sodium phosphate. Or, dissolve 20·3 grammes uranium nitrate in strong acetic acid, and dilute the solution to 1 litre.

Apparatus Required. — Mohr's burette, fitted in a stand, and provided with a Mohr's clip; piece of white porcelain; tripod

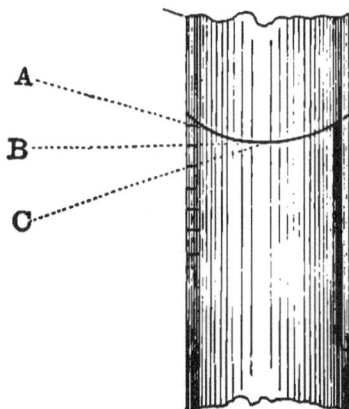

Fig. 19.—Burette Meniscus.   Fig. 20.—Erdmann's Float.

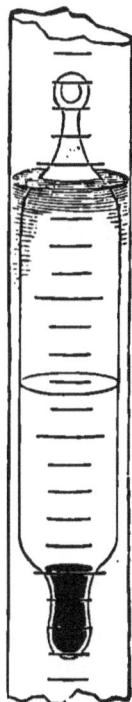

stand and wire-gauze; small beaker; two pipettes, one to deliver 50 cc. and the other for 5 cc. ; glass rod.

13. Reading off the Burette.—In the case of the burette being filled with a watery fluid, note that the upper surface of the water is concave. Always bring the eye to the level of the same horizontal plane as the bottom of the meniscus curve. Fig. 19 shows how different readings may be obtained if the eye is placed at different levels, A, B, C.

14. Erdmann's Float consists of a glass vessel loaded with mercury, so that it will float vertically. It is used to facilitate the reading off of the burette. It has a horizontal line engraved round its middle, and must be of such a width as to allow it just to float freely in the burette. Read off the mark on the burette which coincides with the ring on the float. (Fig. 20.)

15. The **Lime, Magnesia, Iron,** and other inorganic urinary constituents are comparatively unimportant, and have no known clinical significance.

LESSON XV.

ORGANIC CONSTITUENTS OF THE URINE.

1. Urea (CON_2H_4) is the most important organic constituent in urine, and is the chief end-product of the oxidation of the nitrogenous constituents of the tissues and food. It crystallises in silken four-sided prisms, with obliquely cut ends (rhombic system), and when rapidly crystallised, in delicate white needles. It has no effect on litmus; odourless, weak cool-bitter taste, like saltpetre. It is very soluble in water, alcohol, and almost insoluble in ether. It is an isomer of ammonium cyanate. It may be regarded as a biamid of CO_2 or as carbamid $= CO \begin{cases} NH_2. \\ NH_2. \end{cases}$

Urea represents the final stage of the metamorphosis of albuminous substances within the body. More than nine-tenths of all the N. taken in is excreted in the form of urea.

2. Preparation.—Take 20 cc. of fresh filtered human urine, add 20 cc. of baryta mixture—Lesson XVI., 2 (c.)—to precipitate the phosphates. Filter, evaporate the filtrate to dryness in an evaporating chamber, and extract the residue with boiling alcohol. Filter off the alcoholic solution, place some of it on a slide, and allow the crystals of urea, usually long, fine, transparent needles, to separate out. Examine them microscopically.

3. Combinations.—Urea combines with acids, bases, and salts. Evaporate human urine to one-sixth its bulk, and divide the residue into two portions, using one for the preparation or nitrate, and the other for oxalate of urea.

4. Urea Nitrate, CH_4N_2O, HNO_3.

(*a.*) To the concentrated urine add strong, *pure* nitric acid = a precipitate of glancing scales of urea nitrate, which, being almost insoluble in HNO_3, separate out in rhombic plates or six-sided tables, with a mother-of-pearl lustre, and often imbricate arrangement.

(*b.*) Examine the crystals microscopically (Fig. 21).

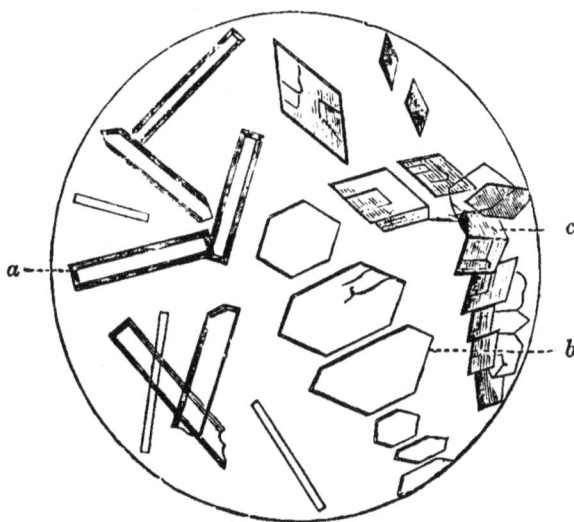

Fig. 21.—*a*, Urea; *b*, hexagonal plates; and *c*, smaller scales, or rhombic plates of urea nitrate.

(*c.*) If only traces of urea are present, concentrate the fluid supposed to contain the urea, place a drop on a slide, put into the drop one end of a thread, apply a cover-glass, and put a drop of *pure* nitric acid on the free end of the thread. The acid will pass into the fluid, and microscopic crystals of urea nitrate will be formed on the thread. After a time examine the preparation microscopically.

5. Urea Oxalate, $(CH_4N_2O)_2 C_2H_2O_4 + H_2O$.

(*a.*) To the other half of the concentrated urine, add a concentrated solution of oxalic acid. After a time crystals of oxalate of urea separate.

(*b.*) Examine them microscopically (Fig. 22).

(*c.*) Add oxalic acid to a concentrated solution of urea = a precipitate of urea oxalate, which may have many forms — rhombic plates, crystalline scales, easily soluble in water.

Fig. 22.—Crystals of oxalate of urea from urine.

(*d.*) Do the same test as described for urea nitrate (**4,** *c.*), but substitute oxalic for the nitric acid.

6. Urea and Mercuric Nitrate $(2U + Hg(NO_3)_2 + 3HgO)$.

(*a.*) To urine or urea solution add mercuric nitrate = a white, cheesy precipitate, a compound of urea and mercuric nitrate. Liebig's method for the estimation of urea is founded on this reaction.

7. Other Reactions of Urea.

Make a strong watery solution of urea, and with it perform the following tests :—

(*a.*) Allow a drop to evaporate on a slide, and examine the crystals which form (Fig. 21, *a*).

(*b.*) Repeat, if you please, the exercises under Lesson XV., 4.

(*c.*) To a strong solution of urea add *pure* nitric acid = a precipitate of urea nitrate (Fig. 21, *b*).

(*d.*) To a strong solution of urea add ordinary nitric acid tinged yellow with *nitrous acid*, or add *nitrous acid* itself; bubbles of gas are given off, consisting of carbon dioxide and nitrogen.

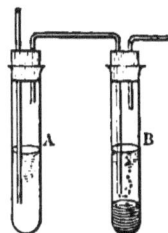

Fig. 23.

(*e.*) Put some of the urea nitrate precipitate obtained in 4 (*a.*) into the test-tube A (Fig. 23), and some lime water in B. Add nitrous acid to A. Cork the tube. The precipitate dissolves. CO_2 and N are given off, the CO_2 makes the lime water in B white.

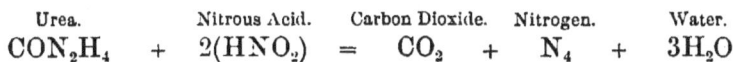

Urea.		Nitrous Acid.		Carbon Dioxide.		Nitrogen.		Water.
CON_2H_4	+	$2(HNO_2)$	=	CO_2	+	N_4	+	$3H_2O$

(*f.*) Mercuric nitrate gives a greyish-white, cheesy precipitate.

(*g.*) Add caustic potash, and heat. The urea is decomposed, and ammonia is evolved.

8. **With Crystals of Urea** perform the following experiments:—

(*a.*) **Biuret Reaction.**—Heat a crystal in a hard tube; the crystal melts, ammonia is given off, and is recognised by its smell and its action on litmus, while a white sublimate of cyanuric acid is deposited on the upper cool part of the tube. Heat the tube until there is no longer an odour of ammonia. Allow the tube to cool, add a drop or two of water to dissolve the residue, and a few drops of caustic soda or potash, and a little very dilute solution of cupric sulphate = a *violet* colour (biuret reaction).

(*b.*) Place a large crystal of urea in a watch-glass, cover it with a saturated watery solution of *furfurol*, and at once add a drop of strong hydrochloric acid, when there occurs a rapid play of colours, beginning with yellow and passing through green, purple, to violet or brown. This test requires care in its performance.

9. **Occurrence.**—Urea occurs in the blood, lymph, chyle, liver, lymph glands, spleen, lungs, brain, saliva, amniotic fluid. The chief seat of its formation is very probably the liver. It also

occurs in the urine of birds, reptiles, and mammalia, but it is most abundant in that of carnivora.

10. Quantity.—An adult excretes 30 to 40 grammes (450 to 600 grs.) daily; a woman less, and children relatively more. It varies, however, with—

(*a.*) *Nature of the Food.*—It increases when the nitrogenous matters are increased in the food, and is diminished by vegetable diet. It is increased by copious draughts of water, salts. It is still excreted during starvation. Muscular exercise has little effect on the amount.

(*b.*) *In Disease.*—In the acute stage of fevers and inflammations there is an increased formation and discharge, also in saccharine diabetes (from the large quantities of food consumed). It is diminished in anæmia, cholera, by the use of morphia, in acute and chronic Bright's disease. If it is retained within the body, it gives rise to uræmia, when it may be excreted by the skin, or be given off by the bowel.

LESSON XVI.

VOLUMETRIC ANALYSIS FOR UREA.

1. Before performing the **volumetric analysis** for urea, do the following reactions, which form the basis of this process :—

(*a.*) To a solution of sodic carbonate, add mercuric nitrate = a yellow precipitate of mercuric hydrate.

(*b.*) To urine, add sodic carbonate, and then mercuric nitrate = first of all a white cheesy precipitate, on adding more mercuric nitrate, a yellow is obtained—*i.e.*, no yellow is obtained until the mercuric nitrate has combined with . the urea, and there is an excess of the mercuric salt.

(*c.*) To urine add *hypobromite of soda*. At once the urea is decomposed and bubbles of gas—N—are given off.

2. Liebig's Volumetric Process for Urea with Sodic Carbonate as

the Indicator.—1 cc. of the SS. (Mercuric Nitrate) = ·01 gramme urea.

(a.) Collect the urine of the twenty-four hours, and measure the quantity.

(b.) If albumin be present separate it by acidification, boiling, and filtration.

(c.) Mix 40 cc. of urine with 20 cc.—i.e., half its volume —of a solution of barium nitrate, and barium hydrate (composed of one volume of solution of barium nitrate and two volumes of barium hydrate both saturated in the cold). This precipitates the phosphates, sulphates, and carbonates.

(d.) Filter through a dry filter to get rid of the above salts. While filtration is going on, fill the burette with the standard solution (SS.) of mercuric nitrate up to the mark 0 on the burette. See that there are no air-bubbles, and that the outflow tube is also filled.

(e.) With a pipette take 15 cc. of the clear filtrate and place it in a beaker. N.B.—This corresponds to 10 cc. of urine. Place a few drops of the sodic carbonate solution (the indicator) on a piece of glass resting on a black background.

(f.) Note the height of the fluid in the burette. Run in the SS. of mercuric nitrate from the burette into the 15 cc. of the mixture, in small quantities at a time, until the precipitate ceases. Stir and mix thoroughly with a glass rod. After each addition, with the glass rod lift out a drop of the mixture and place it on one of the drops of sodic carbonate until a pale yellow colour is obtained. This indicates that all the urea has been precipitated, and that there is an excess of mercuric nitrate. Read off the number of cc. of the SS. used.

(g.) Repeat the experiment with a fresh 15 cc. of the filtrate, but run in the greater part of the requisite SS. at once before testing with sodic carbonate.

Read off the number of cc. of the SS. used, and deduct 2 cc.; multiply by ·01, which gives the amount (in grammes) of urea in 10 cc. of urine.

Example.—Suppose 25 cc. of the SS. were used, and the patient passed 1200 cc. of urine in 24 hours : then, $25 \times \cdot 01 = \cdot 25$ gramme urea in 10 cc.

$$10 : 1200 : : \cdot 25 : x \; . \; \frac{1200 \times \cdot 25}{10} = 30 \text{ grammes of urea in 24}$$

hours.

This method yields approximately accurate results only when the amount of urea is about 2 per cent. With a greater or less percentage of urea, certain modifications have to be made.

3. Correction for Sodic Chloride.—Two cc. were deducted in the above process. Why? On adding mercuric nitrate to a solution containing sodic chloride, the mercuric nitrate is decomposed and mercuric chloride formed, and as long as any sodic chloride is present, there is no free mercuric nitrate to combine with the urea. Proofs of this :

(*a.*) To a solution of sodic chloride (normal saline), add mercuric nitrate = precipitate.

(*b.*) To a solution of sodic chloride (normal saline) add a few crystals of urea, then add mercuric nitrate. At first there is no precipitate, or, if there is, it is redissolved ; but by-and-by a white precipitate is obtained.

(*c.*) To a solution of urea (acid) add mercuric chloride = no precipitate.

4. Solutions Required.

Baryta Mixture.—Prepared as in Lesson XVI., 2, (*c.*)

Mercuric Nitrate Solution (1 cc. = ·01 grm. urea). Dissolve with the aid of gentle heat 77·2 grammes of pure dry oxide of mercury in as small a quantity as possible of HNO_3, evaporate to a syrup, and then dilute with water to 1 litre. A few drops of HNO_3 will dissolve any of the basic salt left undissolved. *N.B.*—The exact strength of this solution must be estimated by titrating it with a standard 2 per cent. solution of urea.

Sodic Carbonate Solution.—20 grains to the ounce of water.

5. Apparatus Required.—Burette fixed in a stand, funnels, beakers, filter paper, glass rod, plate of glass, and three pipettes, 10, 15, and 20 cc.

6. Estimation of Urea by the Hypobromite Method.

The principle of this method rests on the fact that urea is decomposed by alkaline solution of sodic hypobromite. The urea yields CO_2 (which is absorbed by the caustic soda), and N, which is disengaged in bubbles and collected in a suitable apparatus.

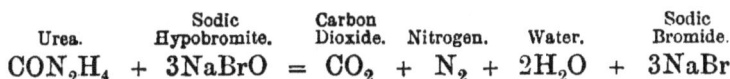

Urea.	Sodic Hypobromite.	Carbon Dioxide.	Nitrogen.	Water.	Sodic Bromide.

$$CON_2H_4 + 3NaBrO = CO_2 + N_2 + 2H_2O + 3NaBr$$

Every 0·1 gramme of urea yields at the ordinary temperature and pressure 37·3 cc. of nitrogen; the calculation, therefore, is simple. Many different forms of apparatus have been devised, including those of Knop and Hüfner, Russel and West, Graham Steele, Simpson, Dupré, Charteris, &c.

(a.) Study these forms of apparatus, but make the experiment with the apparatus of Dupré or Steele.

7. Dupré's Apparatus.—In this apparatus the graduation on the collecting tube represents the percentage of urea, and not cc. of N. The collecting tube, which is clamped above, is placed in a tall vessel containing water, and connected with a small glass flask containing a short test-tube.

(a.) Remove the short test-tube from the flask, and in 'the latter, place 25 cc. of the hypobromite solution.

(b.) With a pipette measure off 5 cc. of the clear filtered urine, and place it in the short test-tube. With a pair of forceps carefully introduce the tube with the urine into the flask, and place the caoutchouc stopper in the latter.

(c.) Test to see if all the connections are tight. Open the clamp at the upper end of the collecting tube, depress the tube in the water until the water inside and outside the tube is at zero of the graduation. Close the clamp, and raise the collecting tube. If the apparatus be tight, no air will pass in, and on lowering the collecting tube the water will stand at zero inside and outside the tube.

7

(*d.*) Mix the urine gradually with the hypobromite solution by gently tilting over the flask, and ultimately move the flask so as to wash out the test-tube with the hypobromite solution. Gas is rapidly given off, the CO_2 is absorbed by the caustic soda, while the N is collected in the graduated measuring tube.

(*e.*) Place the flask in a jar of water at the same temperature as that in the tall jar, and slightly lower the measuring tube. After all effervescence has ceased, and when the N collected in the collecting tube has cooled to the temperature of the room — *i.e.*, in three to five minutes — raise the collecting tube until the fluid inside and outside stands at the same level. Read off the graduated tube; this gives the percentage of urea.

It is to be remembered that other bodies in the urine, such as uric acid (urates) and creatinin— but not hippuric acid — also yield nitrogen by this process; further, that only about 92 per cent. of the N of the urea is given off in the above processes. These sources of fallacy are, however, taken into account in graduating the apparatus.

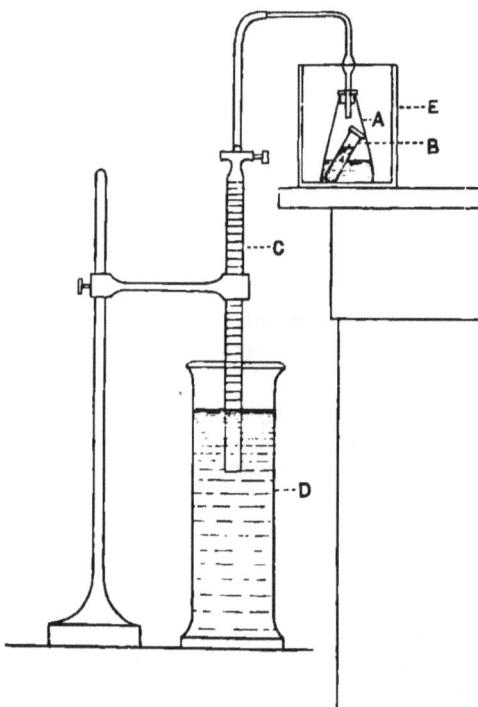

Fig. 24.—G. Steele's apparatus for urea.—A, Flask for hypobromite; B, tube for urine; C, burette; D, vessel with water; E, vessel with water to cool A.

8. Steele's Apparatus.—This is practically the same apparatus,

but a graduated burette is substituted for the graduated collecting tube.

(*a.*) Use this apparatus in a similar manner.

(*b.*) Read off the number of cc. of N evolved, and from this calculate the amount of urea. Every 37·3 cc. N = 0·1 gramme urea.

9. Solutions required.

A. For Dupré's Apparatus.

Hypobromite Solution.—Dissolve 5 cc. of bromine in 45 cc. of a 40 per cent. solution of caustic soda.

N.B.—This solution does not keep, and must be freshly prepared.

B. For Steele's Apparatus.

Hypobromite Solution.—20 grammes of caustic soda are dissolved in water, and the solution is diluted to 250 cc.; after cooling, add 5 cc. of bromine, and mix. Keep in a stoppered bottle in the dark; as it soon decomposes, it should be made fresh.

10. Use also Charteris's apparatus. The bromine and caustic soda are mixed in a marked measure, so that the hypobromite is always fresh, while the collecting tube for the N is so graduated as to indicate a certain percentage of urea.

11. Study Squibb's apparatus. In all these cases directions are supplied with the apparatus.

LESSON XVII.

URIC ACID, URATES, HIPPURIC ACID, CREATININ, &c.

1. Uric Acid ($C_5H_4N_4O_3$) is the constituent of urine in which (next to urea) the most N of the body is excreted, whilst in birds, reptiles, and insects it forms the chief nitrogenous excretion.

2. Quantity.—0·5 gramme (7-10 grs.) daily. The ratio of uric acid to urea is about 1 : 45. It is dibasic, colourless, and crystallises, chiefly in rhombic plates, and when the obtuse angles are rounded the "whetstone" form is obtained. It often crystallises spontaneously in rosettes from saccharine diabetic urine. It is tasteless, reddens litmus, and is very insoluble in water (18,000 parts of cold and 1500 of warm water), insoluble in alcohol and ether. In the urine it occurs chiefly in the form of *acid* urate of soda ($C_5H_2N_4O_3$, HNa) and potash.

(*a.*) In a conical glass, add 5 parts of HCl to 20 parts of urine, put it in a cool place, and allow it to stand for twenty-four hours. Yellow or brownish coloured crystals of uric acid are deposited on the sides of the glass, or form a pellicle on the surface of the fluid like fine grains of cayenne pepper. Both uric acid and its salts (urates) when they occur as sediments in urine are coloured, and the colour is deeper the more coloured the urine. The slow separation of the uric acid is probably due to the presence of phosphatic salts.

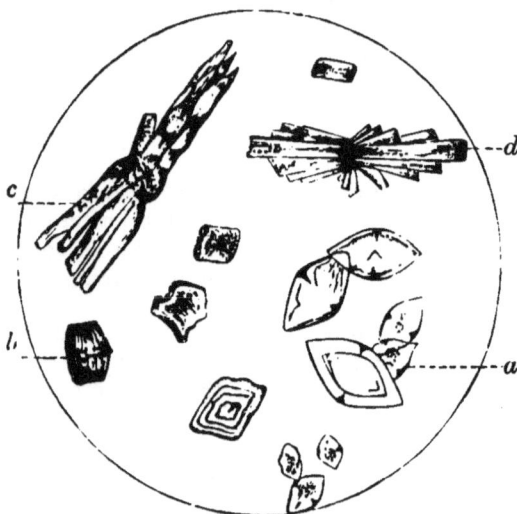

Fig. 25.—Uric acid.—*a*, Rhombic tables (whetstone form); *b*, barrel form; *c*, sheaves; *d*, rosettes of whetstone crystals.

(*b*). Collect some of the crystals and examine them microscopically. The crystals assume many forms, but are chiefly rhombic. They may be whetstone, lozenge-shaped, in rosettes, quadrilateral prisms, &c. They are *yellowish* in colour, although their tint may vary from yellow to red or reddish-brown, depending on the depth of the colour of the urine. (Figs. 25, 26.)

(*c.*) The crystals are soluble in caustic soda or potash. Observe this under the microscope.

(*d.*) Take some serpent's urine—which is solid, and consists chiefly of ammonium urate—dissolve it in a 10 per cent. solution of caustic soda with the aid of heat. Add water, and allow it to stand. Pour off the clear fluid, and precipitate the uric acid with dilute hydrochloric acid. Collect the deposit, and use it for testing.

3. Reactions and Tests.

(*a.*) **Murexide Test.**— Place some uric acid in a porcelain capsule, add nitric acid, and heat gently, taking care that the temperature is not too high — not above 40° C. ‘Very disagreeable fumes are given off, while a yellow or reddish stain remains. Allow it to cool, and bring a rod dipped in ammonia near the stain,

Fig. 26.—Uric acid.—*a*, Rhomboidal, truncated, hexahedral, and laminated crystals; *b*, rhombic prism, horizontally truncated angles of the rhombic prism, imperfect rhombic prisms ; *c*, prism with a hexahedral basic surface, barrel-shaped figure, prism with a hexahedral basal surface ; *d*, cylindrical figure, stellate and superimposed groups of crystals.

or moisten it with strong ammonia, when a *purple-red* colour of *murexide*, $C_8H_8(NH_4)N_5O_6$, appears. It turns violet on adding caustic potash.

(*b.*) Repeat the experiment, but act on the residue with caustic soda or potash, when a violet-blue colour—discharged by heat—is obtained. When uric acid is acted on by nitric acid, alloxantin ($C_8H_4N_4O_7$) is formed, which on being further heated yields alloxan ($C_4H_2N_2O_4$), which strikes a purple colour—murexide—with ammonia.

(*c.*) Place some uric acid on a microscopic slide, and

dissolve it in liquor potassæ. Heat if necessary; add hydrochloric or nitric acid just to excess, and examine with the microscope the crystals of uric acid which form. They may be transparent rhombs with obtuse angles, dumb-bells, or in rosettes.

(*d.*) Dissolve some uric acid in caustic soda, add a drop or two of Fehling's solution—or dilute cupric sulphate and caustic soda—and boil = a white precipitate of cupric urate, which after a time becomes greenish.

(*e.*) **Schiff's Test.**—Dissolve some uric acid in a small quantity of sodic carbonate. Place, by means of a glass rod, a drop of solution of silver nitrate on filter-paper, and on this place a drop of the uric acid solution. A dark brown or black spot of reduced silver appears.

(*f.*) **Garrod's Microscopic Test.**—Add 6 to 8 drops of glacial acetic acid to 5 cc. urine in a watch-glass, put into it a few silk threads, and allow the whole to stand for twenty-four hours, taking care to prevent evaporation by covering it with another watch-glass or small beaker. Examine the threads microscopically for the characteristic crystals of uric acid, which are soluble in KHO. A similar reaction may be done on a microscopic slide.

(*g.*) Heat some uric acid in a test-tube. It blackens and gives off the smell of burnt feathers.

4. Uric Acid Salts (Urates).—Uric acid forms salts (chiefly acid), with various bases, which are soluble with difficulty in cold, but readily soluble in warm water. HCl and acetic acid decompose urates, when the uric acid crystallises.

Urates form one of the commonest and least important deposits in urine. There is usually a copious precipitate, varying in colour from a light pink or brick-red to purple. They occur in catarrhal affections of the intestinal canal, after a debauch, in various diseases of the liver, in rheumatic and feverish conditions. They frequently occur as the "milky" deposit in the urine of children. Urates constitute the "lateritious" deposit, or "critical" deposit of the older writers. Urates frequently occur even in health, especially when the skin is very active (in summer), or after severe muscular exercise; when much water is given off by the skin, and a small quantity by the kidneys.

When the urine is passed it is quite clear, but on standing for a time it becomes turbid, and a copious reddish-yellow—sometimes like pea soup—or purplish precipitate occurs, because urates are more soluble in warm water than in cold; and when there is only a small quantity of water to hold the urates in solution, on the urine cooling they are precipitated. Their occurrence is favoured by an acid reaction, a concentrated condition of the urine, and a low temperature.

The urates deposited in urine consist chiefly of sodic urate, mixed with a small amount of ammonium urate.

5. Tests.—Secure a specimen of " urates " in urine.

(*a.*) Observe the naked-eye characters. The deposit is usually copious = yellowish-pink, reddish, or even shading into purple. The deposit moves freely on moving the vessel, and its upper border is fairly well defined.

(*b.*) Place some in a test-tube. Heat gently the upper stratum. It becomes clear, and on heating the whole mass of fluid, it also becomes clear, as the urates are dissolved by the warm liquid.

(*c.*) Place some of the deposit on a glass slide, add a drop of hydrochloric acid, and uric acid is deposited in one or more of its many crystalline forms. Examine the crystals microscopically.

(*d.*) Examine the deposit microscopically. The urates are usually " amorphous," but the urate of soda may occur in the form of small spheres covered with spines, and the ammonium urate, of spherules often united together (Fig. 17, *b*).

(*e.*) Make a saturated solution of uric acid in caustic soda. Place a drop of the mixture on a slide, allow it to evaporate. Examine it microscopically, when the urate of soda in the form of spheres covered with spines will be obtained. (Fig. 27.)

Fig. 27.—Urate of soda.

(*f.*) The same result as in (*e.*) is obtained by dissolving the ordinary deposit of urates with caustic soda, and allowing some of it to evaporate on a slide.

6. Hippuric Acid, $C_9H_9NO_3$.—This substance is so called because it occurs in large quantity in the urine of the horse and many herbivora, chiefly in the form of alkaline hippurates.

Quantity in man ·5 to 1 gramme daily. It is a conjugate acid, which, when boiled with alkalies and acids, takes up water and splits into benzoic acid and glycocin. It occurs in colourless four-sided prisms, usually with two or four bevelled surfaces at their ends. It has a bitter taste. Benzoic acid, oil of bitter almonds, benzamid, cinnamic acid, and toluol reappear in the urine as hippuric acid. The benzoic acid unites with the elements of glycocoll (glycin), and is excreted as hippuric acid in the urine.

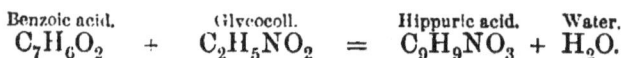

Benzoic acid.		Glycocoll.		Hippuric acid.		Water.
$C_7H_6O_2$	$+$	$C_2H_5NO_2$	$=$	$C_9H_9NO_3$	$+$	H_2O.

The amount is *increased* by eating pears, apples with their skins, cranberries, and plums. Nothing is known of its clinical significance. It seems to be formed chiefly from the husks or cuticular structures.

Tests and Reactions.

(*a.*) Heat some crystals in a dry tube. Oily red drops are deposited in the tube, while a sublimate of benzoic acid and

Fig. 28.—Hippuric Acid.

ammonium benzoate are given off. The latter is decomposed, giving the odour of ammonia, while there is an aromatic odour of oil of bitter almonds.

(*b.*) Examine the colourless four-sided prisms with the microscope (Fig. 28).

7. Preparation of Hippuric Acid.—Take 100 cc. of cow's urine, and evaporate it to one half its bulk ; add hydrochloric acid, and set it aside. The brown mass is collected, dried between folds of blotting-paper, re-dissolved in a very small quantity of water, and mixed with charcoal, then filtered and set aside to crystallise.

8. Creatinin ($C_4H_7N_3O$) is a derivative of the creatin of muscle. If creatin be boiled with acids or with water for a long time, it loses water, and becomes converted into a strong base—creatinin.

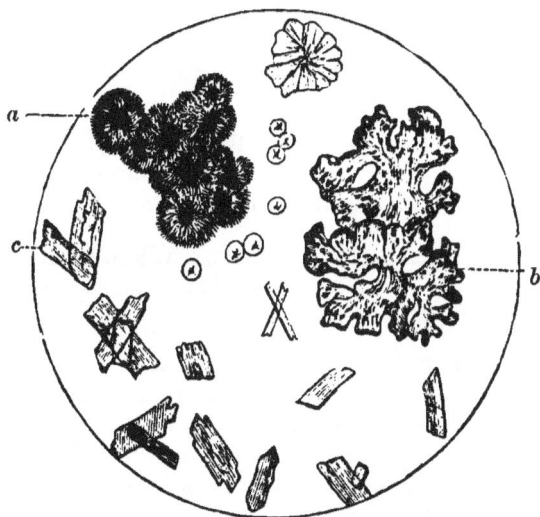

Fig. 29.—Creatinin-zinc-chloride.—*a*, Balls with radiating marks; *b*, crystallised from water ; *c*, rarer forms from an alcoholic extract.

Quantity, 0·5 to 1 gramme (7 to 15 grs.) It is easily soluble in water and alcohol, and forms colourless oblique rhombic crystals. It unites with acids, and also with salts, chiefly with $ZnCl_2$; the creatinin-zinc-chloride is used as a microscopic test for its presence. It rarely occurs as a deposit, and nothing is known of its clinical significance.

9. Preparation of Creatinin.—Take 250 cc. of urine, precipitate
it with milk of lime, and filter.   Evaporate the filtrate to a
syrupy consistence, and extract it with alcohol.   Filter, and to
the filtrate add a drop or two of a neutral solution of zinc chloride, and set the vessel aside.   After a time creatinin-zinc-chloride is deposited on the sides of the vessel.

10. Tests and Reactions.

(*a.*) Examine the deposit **microscopically.**   It forms round
brownish balls, with radiating lines (Fig. 29).

(*b.*) **Weyl's Test.**—To urine add a very dilute solution of
sodic nitro-prusside, and very cautiously caustic soda = a
ruby-red colour, which is evanescent, passing into a straw
colour.

11. Colouring-Matters of the Urine.

Several colouring-matters seem to be present in the urine,
including :—

(1.) **Urobilin,** which occurs especially in the high-coloured urines·
of fever.   It gives urine its red or reddish-yellow colour, and
on the addition of ammonia it becomes yellow.

(2.) **Uro-Chrome** is a yellow pigment.   When a watery solution
is exposed to the air it oxidises, and becomes red owing to the
formation of uro-erythrin.

(3.) **Indigo-forming Substance (Indican).**—This is derived from
indol, C_8H_7N, which is developed in the intestinal canal from the
pancreatic digestion of proteids, and also from the putrefaction of
albuminous bodies.   It may also be formed from bilirubin.   In
urine it is a yellow pigment, and is more plentiful in the urine
of the dog and horse.

12. General Reactions.

(*a.*) Add to normal urine a quarter of its volume of HCl,
and boil = a fine pink or yellow colour.

(*b.*) Add nitric acid = a yellowish-red colour, usually
deeper than the original colour.

(*c.*) To two volumes of sulphuric acid in a test-tube, add

one of urine, but drop the latter from a height. The mixture becomes more or less garnet red if indican be present.

(*d.*) Add acetate of lead = a precipitate of chloride, sulphate, and phosphate of lead. Filter, the filtrate is an almost colourless solution. This substance is used to decolourise urine for the saccharimeter.

(*e.*) Filter urine through animal charcoal, the urine will be decolourised.

If possible, obtain a dark-yellow coloured urine, and perform the following test :—

(*f.*) Take 40 drops of urine + 3 to 4 cc. of strong HCl and 2 to 3 drops of HNO_3; on heating, a *violet red* colour with the formation of true rhombic crystals of indigo-blue, indicates the presence of indican.

The **pathological pigments**—bile, blood, &c.—occurring in urine will be referred to later.

13. Mucus.—A trace of mucus occurs normally in urine. Collect fresh urine in a tall vessel, and allow it to stand for some time, when fine clouds ("mucous clouds") like delicate cotton-wool appear. These consist of mucus entangling a few epithelial scales.

(*a.*) If urine contain an excess of mucus, on adding a saturated solution of citric acid to form a layer at the bottom of the test-tube, a haziness at the line of junction of the urine and acid indicates mucus. There is no deposit with healthy, freshly-passed urine. Citric acid is used because it is heavier than acetic.

LESSON XVIII.

ABNORMAL CONSTITUENTS OF THE URINE.

Some of the substances referred to in the subsequent lessons are present in excessively minute traces in normal urine—*e.g.*, sugar; and in the urine of a certain percentage of persons *apparently* enjoying perfect health, minute traces of albumin are sometimes present. When, however, these substances occur in considerable quantity, then their presence is of the utmost practical and diagnostic value, and is distinctly abnormal. It is quite certain that serum-albumin is never found in any considerable amount in normal urine.

1. **Albumin in Urine.**—When albumin occurs in notable quantity in the urine, it gives rise to the condition known as **albuminuria.**

Various forms of proteid bodies may occur in the urine. The chief one is **serum-albumin;** but, in addition, **serum-globulin, hemi-albumose, peptone, acid-albumin,** and **fibrin** may be found.

2. **Tests.**—The demonstrator will procure an albuminous urine. With it perform the following tests. In every case the urine must be clear before testing, which can be secured by careful filtration.

(*a.*) **Coagulation by Heat.**—Place 10 cc. of urine in a test-tube and boil. Near the boiling point, if albumin be present in small amount, it will give a haziness; if in a large amount, a distinct coagulum. On standing, the coagulum is deposited. **Precautions.**—(i.) Always test the reaction of the urine, for albumin is only precipitated by boiling in a neutral or acid medium. Hence if the urine be alkaline, boiling will not precipitate any albumin that may be present. (ii.) Boil the upper stratum of the fluid first of all, holding the tube obliquely, taking care that the coagulum does not stick to the glass, else the tube is liable to break. (iii.) Heat, by driving off the CO_2, also precipitates *earthy phosphates* if they are present in large amount, hence a turbidity on boiling is not sufficient proof of the presence of albumin. The points of distinction are, that albumin goes down before the boiling point is reached (coagulated at 75° C.), while phosphates are precipitated at the boiling point. Again, the phosphatic deposit is soluble in an acid— *e.g.,* acetic or nitric—while the albuminous coagulum is insoluble in these fluids. Some, therefore, advise that the test be done in the following manner :—

(*b.*) Acidulate the urine with a few drops of acetic or nitric acid, and then boil. If nitric acid be used, add one-tenth to one-twentieth of the volume of urine. **Precautions.**—If the urine contain only very minute traces of albumin, the latter may not be precipitated if too much nitric acid be added, as the acid-albumin is kept in solution. If too little acid be added, the albumin may not be precipitated, as only a part of the basic phosphates are changed into acid phosphates, and the albumin remains in solution as an albuminate (a compound of the albumin with the base). On heating the urine of a person who is taking copaiba, a deposit may be obtained, but its solubility in alcohol at once distinguishes it from coagulated albumin. This test acts with serum-albumin, and globulin, and if the deposit occurs only after cooling, also with albumose, but not with peptone.

(*c.*) Acidulate 10 cc. of the urine with acetic acid, add one fifth of its bulk of a saturated solution of magnesic or sodic sulphate, and boil = a precipitate.

(*d.*) **Heller's Cold Nitric Acid Test.**—Take a conical test-glass, and place in it 15 cc. of the urine. Incline it, and pour slowly down its side strong nitric acid = a white cloud at the line of junction of the fluids. **Precautions.**—A *crystalline* deposit of urea nitrate is sometimes, though very rarely, obtained with a very concentrated urine. If the urine contain a large amount of urates, they may be deposited by the acid, but the deposit in this case occurs above the line of junction, and disappears on heating. It is not obtained if the urine be diluted beforehand.

(*e.*) **Acetic Acid and Potassium Ferrocyanide.**—Acidify strongly with acetic acid, and add a solution of potassium ferrocyanide = a white precipitate, varying in amount with the albumin present. The reaction may be done as follows:—Mix a few cc. of moderately strong acetic acid with some solution of potassic ferrocyanide, and pour this over some urine in a test-tube by the contact method (*d.*) The presence of albumin is indicated by a white deposit in the form of a ring at the line of junction of the fluids. A solution of platino-potassic cyanide may be used instead of the ferrocyanide. The solution of the former is colourless. This test precipitates serum-albumin, globulin, albumose, but not peptone.

(*f.*) **Picric Acid.**—Use a saturated watery solution, and apply it by the contact-method of Heller (*d.*)   The urine is below, and the picric acid on the top.   A rapidly-formed deposit at the line of junction of the fluids indicates the presence of a proteid ; the deposit is not dissolved by heat.   *N.B.*—Picric acid precipitates all the forms of proteid which occur in urine.   It also precipitates mucin, but in this case the deposit usually forms slowly, and after a time.   If a person be taking quinine, a haziness is obtained in the urine on adding picric acid, but it disappears on heating.   Dr. Johnson and Professor Grainger Stewart recommend it as one of the most reliable tests for albumin we possess.

(*g.*) Do the **biuret reaction**, which reacts with albumin, albumose, globulin, and peptone.

(*h.*) **Metaphosphoric Acid** completely precipitates albumin, but it must be freshly prepared, and is difficult to keep. Hence it is not satisfactory.

(*i.*) **Acidulated Brine,** as suggested by Roberts, consisting of a saturated solution of sodic chloride with 5 per cent. of dilute hydrochloric acid (B.P.), may be used, but it sometimes gives a precipitate with normal urine.   Nor is **potassio-mercuric iodide** satisfactory.   In cases of doubt use several tests, especially **2** (*b.*), (*c.*), (*e.*), and (*f.*)

3. Dry Tests.

(*a.*) Use the ferrocyanic pellets introduced by Dr. Pavy.

(*b.*) Use the test-papers—citric acid and ferrocyanide of potassium—introduced by Dr. Oliver.

4. Globulinuria.—Serum-globulin is present in nearly every albuminous urine.   Procure such a urine.   It gives the reactions described under **2.**

(*a.*) Take a tall glass and fill it with water.   Drop the urine into the water, and observe if a milkiness is seen in the water, indicating the presence of a globulin.   This body is not soluble in pure water, but in weak saline solutions (Lesson I., 6), hence on diluting the urine it is precipitated.

(*b.*) Test the urine by the contact method with a saturated solution of magnesic sulphate.

(*c.*) If globulin be present along with serum-albumin, make the urine alkaline with ammonia, allow it to stand for an hour, filter, and to the filtrate add an equal volume of a saturated solution of ammonia sulphate. A white flocculent precipitate indicates globulin.

5. **Albumosuria.**—Albumose, which, however, is really a mixture of four different proteids, has been found in cases of osteomalacia. If such a urine can be procured, do test 2 (*b.*), using nitric acid ; the deposit only takes place after a long time or on cooling, and in fact the urine sometimes becomes almost solid, but is dissolved by heat. If there is a deposit, filter and test the filtrate for proteid reactions—*e.g.*, the biuret test. It will give a precipitate with acetic acid and potassic ferrocyanide. Then saturate a portion of the urine with sodic chloride, and acidify with acetic acid = a precipitate, which dissolves on adding much acetic acid and heating, and disappears on cooling.

6. **Peptonuria.**—Peptone is frequently present in albuminous urine. Peptone is most frequently present in urine in cases where there is an accumulation and breaking up of leucocytes or pus corpuscles, as in the stage of resolution of pneumonia, suppurative processes, and in other diseases. Procure such a urine. It is well to get rid of the albumin by acidification with acetic acid and boiling.

(*a.*) Put some urine in a test-tube, and by the contact method pour on some Fehling's solution. At the line of junction a phosphatic cloud is formed, and, if peptones be present, above it a rose-pink colour. If albumins also be present, a violet colour is obtained. Hemialbumose gives the same reaction.

(*b.*) Test the urine with tests 2 (*b.*) and (*c.*), and if they are negative, and acetic acid alone gives a turbidity, suspect peptone. Test a portion with acetic acid and phospho-wolframic acid mixed with acetic acid. If peptone be present, there is either immediately or after standing some time a deposit. If there be no deposit on standing, peptone is absent. It is better to precipitate the mucin with neutral lead acetate, and then to apply the above test.

7. **Quantitative Estimation of Albumin.**—This can only be done accurately by precipitating the albumin, drying and weighing it ;

but as this is a tedious process and requires much time, it is not suitable for the physician.

8. Esbach's Method (Albumimeter).

A. The Reagent.—Dissolve 10 grammes of picric acid, and 20 grammes of citric acid in 800 cc. of boiling water, and make up the solution to a litre.

Dr. Johnson finds that a solution of picric acid in boiling water (5 grains to the ounce) gives the same result.

B. Process.—Pour urine into the tube (6 in. × $\frac{5}{8}$ inch) up to the mark U, then the reagent up to the mark R, mix thoroughly. Set the tube aside for twenty-four hours, and then read off on the scale the height of the coagulum. The figures indicate the grammes of dried albumin in a litre of urine—i.e., the percentage is obtained by dividing by ten. If the coagulum is above 4, dilute the urine first with one or two volumes of water, and then multiply the resulting figure by 2 or 3 as the case may be. If the urine be alkaline, it must first be acidulated by acetic acid. If the amount of albumin be less than 0·5 grammes per litre, it cannot be accurately estimated by this method.

LESSON XIX.

BLOOD, BILE, AND SUGAR IN URINE.

1. Blood in Urine (Hæmaturia).—The blood may come from any part of the urinary apparatus. If from the *kidney*, it is usually small in amount and well mixed with the urine, and the microscope may reveal the presence of "blood-casts," i.e., blood-moulds of the renal tubules. Large coagula are never found, and the urine not unfrequently is "smoky." From the *bladder* or *urethra*, usually the urine is bright red, and relatively large coagula are frequently present. In all forms, blood-corpuscles are to be detected by the microscope, and albumin by its tests.

(a.) Examine the naked-eye characters of a specimen. It may be any tint from red to brown, but if the blood is well mixed with the urine, the latter usually has a "smoky" appearance.

(*b.*) Collect any deposit and examine it microscopically for blood-corpuscles, which, however, are frequently discoloured or misshapen.

(*c.*) Examine for the spectrum of oxy-hæmoglobin or met-hæmoglobin " (Lesson IV.)

(*d.*) **Heller's Blood Test.**—Make the urine strongly alkaline with caustic soda, and boil. On standing, a deposit of earthy phosphates coloured red or brown by hæmatin occurs, the deposit carrying down the altered colouring-matter of the blood with it. This is not a satisfactory test.

(*e*). Mix some freshly prepared tincture of guaiacum with urine, and pour on it some ozonic ether ; a blue colour indicates the presence of hæmoglobin.

(*f.*) The urine gives the reactions of albumin.

2. Hæmoglobinuria is applied to that condition where hæmoglobin is excreted through the kidney as such, and is not contained within blood-corpuscles. The urine contains hæmoglobin, but not the blood-corpuscles as such. It occurs when blood-corpuscles are destroyed within the blood-vessels, as after the transfusion of the blood of one species into the blood-vessels of another species ; after the transfusion of warm water ; the injection of a solution of hæmoglobin into a vein ; and after extensive destruction of the skin by burning. It also occurs in purpura, scurvy, often in typhus or scarlet fever, pernicious malaria, in " periodic ,hæmoglobinuria," and after the inhalation of arseniuretted hydrogen.

(*a.*) The urine gives the same reactions as in hæmaturia, but no blood-corpuscles are detected by the microscope.

3. Bile in Urine.—The biliary constituents appear in the urine in cases of jaundice and in poisoning with phosphorus. One may test for the *bile-pigments,* or the *bile-acids,* or both.

A. Bile-Pigments.

(*a.*) **Colour.**—The urine has usually a yellow or yellowish-green colour, and it froths very easily when shaken. Filter-paper dipped into it gives a yellow stain on drying.

(*b.*) **Gmelin's Test** (Nitric acid containing nitrous acid).— (1) Place a few drops of the suspected urine on a white porcelain plate, and near it a few drops of the impure nitric acid; let the fluids run together and the usual play of colours is observed (Lesson IX., 6). (2) Take urine in a test-tube, pour in the *impure* HNO_3, until it forms a stratum at the bottom ; if bile-pigments be present, at the line of junction of the fluids a play of colours takes place— from above downwards —green, blue, violet or dirty red, and yellow. Nearly all urines give a play of colours, but *green* is the necessary and characteristic colour to prove the presence of bile-pigments. Or (3) filter the urine several times through the same filter, dry the filter-paper, and to it apply the impure nitric acid, when the same play of colours is observed.

(*c.*) A solution of methyl-violet poured on icteric urine by the contact method gives a bright carmine ring at the point of contact.

(*d.*) If much bile-pigment be present, the following test succeeds :—Mix the urine with caustic potash (1 KHO to 3 water), and add hydrochloric acid. The fluid becomes green, due to the formation of biliverdin.

B. **Bile-Acids** (Glyco-cholic, and Tauro-cholic acids).

(*a.*) **Pettenkofer's Test.**—Add to urine a few drops of syrup of cane sugar (8 per cent.), mix them, and pour strong sulphuric acid down the side of the tube until it forms a layer at the bottom. The temperature must not rise above 70° C., nor must the urine contain albumin. At the line of junction a *cherry-red* or *purple-violet* colour indicates the presence of the bile-acids. Or proceed as follows:—Shake the tube with the urine and the syrup to get a froth, and when the sulphuric acid is added, the froth shows the colour. *N.B.*—The test in this simple form often fails with urine, and in fact there is no satisfactory simple test for minute quantities of these acids in urine.

(*b.*) **Strasburg's Modification.**—Dissolve cane sugar in the suspected urine, dip into it filter-paper, and allow it to dry. Touch the paper with a glass rod dipped in strong sulphuric acid, a purple-violet colour indicates the presence of the bile-pigments.

(*c.*) Try the sulphur test (Lesson IX., 5).

4. Sugar in Urine (Glycosuria).—Brücke maintains that the merest trace of *glucose* or *grape sugar* is normally present in urine. In *Diabetes mellitus*, however, it occurs in considerable amount, and is, of course, then quite abnormal.

Characters of Diabetic Urine.

(1) The patient usually passes a very large *quantity* of urine even to 10,000 cc., and although the quantity of fluid is large,

(2) The *specific gravity* is high—1030 to 1045—due to the presence of the grape sugar.

(3) The *colour* is usually a very pale straw, from the dilution—not diminution—of the urine pigments. The urine is often somewhat turbid.

(4) It has a heavy sweet *smell*, and usually froths when poured from one vessel into another.

N.B.—When the quantity of urine is above normal, and the specific gravity reaches 1030, suspect the presence of grape sugar.

5. Tests.—In all cases remove any albumin present.

(*a.*) **Moore's Test.**—To urine add an equal volume of caustic soda or potash, and boil the upper stratum of the fluid. If much sugar be present, a dark sherry or bistre-brown colour is obtained. The colour may vary from a light yellow to a dark brown (due to the formation of glucic and melassic acids), according to the amount of sugar present. This is not a delicate test.

(*b.*) **Trommer's Test.**—Add to the urine one-third its bulk of caustic soda solution, and then a *few drops* of a solution of cupric sulphate, and a clear blue solution of the hydrated oxide is obtained. Boil the *upper* stratum of the fluid. If sugar be present, a yellow or yellowish-red ring of reduced cuprous oxide is obtained.

(*c.*) **Fehling's solution** is alkaline potassio-tartarate of copper ($K_2Cu2C_4H_4O_6$). Place some Fehling's solution in a test-tube and boil it. If no discoloration (yellow) takes place, it is in good condition. Add a few drops of the suspected urine and boil, when the mixture suddenly turns to an opaque yellow or red colour, which indicates the presence of a reducing sugar.

(*d.*) **Böttger's Test.**—Mix the urine with an equal volume of sodic carbonate solution, add a little basic bismuthic nitrate, and boil for a short time. A grey or black deposit indicates the presence of a reducing sugar.

(*e.*) **Picric Acid.**—To the urine add an equal volume of a saturated watery solution of picric acid, and then caustic potash. Boil, an intensely deep red or reddish-brown colour indicates the presence of a reducing sugar. The larger the amount of sugar, the deeper the tint. The coloration is due to the formation of picramic acid.

(*f.*) **Indigo-Carmine Test.**—To the urine add sodic carbonate solution and indigo-carmine solution until a blue colour appears. Boil, and a yellow colour is obtained, if sugar be present, owing to the reduction of indigo-blue to indigo-white. Pour the fluid into a cold test-tube, when the blue colour is restored, a beautiful play of colours intervening between the yellow and the blue. This is not a satisfactory test.

6. Preparation of Fehling's Solution.—34·64 grammes of pure crystalline cupric sulphate are powdered and dissolved in 200 cc. of distilled water; in another vessel dissolve 173 grammes of Rochelle salts in 480 cc. of pure caustic soda, specific gravity 1·14. Mix the two solutions, and dilute the deep coloured fluid which results to 1 litre. It is better to keep the two solutions separate in stoppered bottles, and mix them as required.

N.B.—Fehling's solution ought not to be kept too long; it is apt to decompose, and should therefore be kept away from the light, or protected with opaque paper pasted on the bottle. Some other substances in urine—*e.g.*, uric acid—reduce cupric oxide. *In all cases see that there is an excess of the test present.*

LESSON XX.

QUANTITATIVE ESTIMATION OF SUGAR.

1. By the Saccharimeter.

Study the use of some form of saccharimeter. The portable form made by Zeiss is very convenient. A coloured urine must first be decolourised by acetate of lead [Lesson XVII., 12 (d.)].

2. Volumetric Analysis by Fehling's Solution.—10 cc. of Fehling's solution = ·05 grms. of sugar.

(a.) Ascertain the quantity of urine passed in twenty-four hours.

(b.) Filter the urine, and remove any albumin present by boiling and filtration.

(c.) Dilute 10 cc. of Fehling's solution with about five to ten times its volume of distilled water, and place it in a white porcelain capsule on a wire gauze support under a burette. [It is diluted because any change of colour is more easily observed.]

(d.) Take 5 cc. of the diabetic urine, add 95 cc. of distilled water, and place the *diluted* urine in a burette.

(e.) Boil the diluted Fehling's solution, and whilst it is boiling, gradually add the diluted urine from the burette, until all the cuprous oxide is precipitated as a reddish powder, and the supernatant fluid has a straw-yellow colour, not a trace of blue remaining. This is best seen when the capsule is tilted. It is not advisable to spend too much time in determining when the blue colour disappears, as it is apt to return on cooling.

(f.) Read off the number of cc. of *dilute* urine employed. If 36 cc. were used, this, of course, would represent 1·8 cc. of the original urine.

(g.) Make a second determination, using the data of the first, and in this case run in at once a little less of the dilute urine than was required at first.

Example.—Suppose the patient passes 8550 cc. of urine, then as 1·8 cc. of urine reduced all the cupric oxide in the 10 cc. of Fehling's solution, it must contain ·05 gramme sugar; hence

$$1·8 : 8550 : : ·05 \therefore \frac{8550 \times ·05}{1·8} = 237·5 \text{ grammes of sugar passed}$$

in twenty-four hours.

3. Picro-Saccharimeter of G. Johnson.

Solutions Required.

(1.) A solution of ferric acetate equal to that yielded by a solution of sugar containing ¼ grain per fluid ounce.

(2.) Saturated solution of picric acid.

(3.) Liquor Potassæ B.P.)

Fig. 30.—Picro-saccharimeter.

(a.) Measure 1 fluid drachm of urine into the boiling tube, add 30 minims of liquor potassæ and 80 minims of the saturated solution of picric acid. Make up to the 4-drachm mark on the tube with distilled water. Boil for one minute.

(b.) Dip the tube in cold water to cool it. The volume must be exactly 4 drachms. If it is less, add water; if more, evaporate it. If the colour of the boiled liquid is the same as that of the ferric acetate ¼ grain standard, or paler, the urine contains 1 grain of sugar per fluid ounce, or less.

(c.) Should the colour be *darker* than the standard, place some of the boiled liquid into the graduated stoppered tube (Fig. 30) to fill 10 divisions of the scale, while the stoppered tube affixed to the former is filled with the SS. of ferric acetate. Fill up the graduated tube with distilled water until the dark red liquid has the same colour as that of the SS. These tints are best compared in the flat-bottomed tubes supplied with the apparatus.

(d.) Read off the level of the fluid in the

saccharimeter, each division above 10 = 0·1 grain per fluid oz. Thus, 13 divisions = 1·3 grains per fluid oz.

(*e.*) If more than 8 grains per oz. are present, further dilution is required. Full instructions are supplied with the apparatus.

4. Aceto-Acetic Acid is found in certain diabetic urines, but not in all.

(*a.*) To the urine add ferric chloride; a *red* colour is obtained if this acid be present. If there is a deposit of phosphates, filter. The colour disappears on heating.

If a diabetic urine containing aceto-acetic acid be distilled, this acid is decomposed, and **aceton** is obtained.

5. Tests for Aceton.

(*a.*) **Lieben's Test.**—To a weak, watery solution of aceton add solution of iodine dissolved with the aid of potassic iodide, and then caustic soda. A yellow precipitate of iodoform is obtained. The precipitate is generally described as forming hexagonal plates or radiate stars, but I have generally found it to be amorphous or granular. Other substances give the iodoform reaction.

(*b.*) Smell the peculiar ethereal odour of aceton.

(*c.*) **Legal's Test.**—Add caustic soda solution, and then a solution of freshly-prepared sodium nitro-prusside, and acetic acid = a red colour.

In all cases employ both tests, but they only give a decided reaction in urine when the aceton is in considerable amount. To be quite certain that aceton is present, a considerable amount of the urine must be distilled, and the tests applied to the distillate.

6. Tests for Phenol.—To a watery solution of phenol

(*a.*) Add ferric chloride = a bluish-violet colour.

(*b.*) Add bromine water = a yellow (or rather white) precipitate of bromine compounds.

(c.) Add Millons' reagent = a beautiful *red* colour or deposit. This reaction is aided by heat.

7. Pyrocatechin is sometimes found in urine. The method of obtaining it requires too much time to be done in this course.

Tests.

(*a.*) To a dilute solution add ferric chloride = a *green* colour, which becomes *violet* on the addition of sodic bicarbonate.

(*b.*) Add ammonia and silver nitrate, which give a *black* precipitate of reduced silver.

LESSON XXI.

URINARY DEPOSITS—CALCULI AND GENERAL EXAMINATION OF THE URINE.

1. Mode of Collecting Urinary Deposits.—Place the urine in a conical glass, cover it, and allow it to stand for twelve hours. Note the reaction before and after standing. With a pipette remove some of the deposit and examine it microscopically. There are two classes of deposits, *organised* and *unorganised.*

ORGANISED DEPOSITS.

1. Pus (p. 120).
2. Blood (p. 112).
3. Epithelium.
4. Renal tube casts.

5. Spermatozoa.
6. Micro-organisms.
7. Elements of morbid growths and entozoa.

2. Pus in Urine (Pyuria) produces a thick creamy yellowish-white sediment after standing, although its appearance varies with the reaction of the urine. If the urine be *acid*, the precipitate is loose, and the pus corpuscles discrete; if *alkaline*, and especially from ammonia, it forms a thick, tough, glairy mass. The urine is usually alkaline, and is always albuminous, and rapidly undergoes decomposition. Pus is found in the urine in leucorrhœa in the female, gonorrhœa, gleet, cystitis, pyelitis, from bursting of an abscess into any part of the urinary tract, &c.

(*a*). Donné's test.—Filter off the fluid, and add to the deposit a small piece of caustic potash, or a few drops of strong solution of caustic potash; the deposit becomes ropy and gelatinous, and cannot be dropped from one vessel into another—due to the formation of alkali-albumin—the deposit is pus. The same reagent with *mucus* causes the deposit to become more fluid and limpid, to clear up, and look like unboiled white of egg.

(*b*.) With the microscope numerous pus corpuscles are seen, which when acted on by acetic acid show a bi- or tri-partite nucleus. This test is not absolutely conclusive.

(*c*.) Urine containing pus gives the reactions for albumin, while, if mucus alone be present, it gives only those for mucin.

UNORGANISED DEPOSITS.

A. In Acid Urine.

1. *Amorphous.*

(*a*.) **Urates.**—Soluble when heated, redeposited in the cold; when hydrochloric acid is added microscopic crystals of uric acid are formed = urates.

(*b*.) **Tribasic Phosphate of Lime.**—Not dissolved by heat, but disappears without efferves-cence on adding acetic acid. It is probably tribasic phosphate of lime ($Ca_3 2PO_4$).

(*c*.) **Oil Globules.**—Very small highly refractive globules, soluble in ether (very rare).

2. *Crystalline.*

(*a*.) **Uric Acid.**—Recognised by the shape and colour of the crystals and their solubility in KHO.

(*b*.) **Oxalate of Lime.**—Octa-hedral crystals, insoluble in acetic acid (Fig. 31, *b*, *c*).

B. In Alkaline Urine.

1. *Amorphous.*

(*a*.) **Tribasic Phosphate of Lime** dissolves in acids without effervescence.

(*b*.) **Carbonate of Lime.** (See (*c*.) below.)

2. *Crystalline.*

(*a*.) **Triple Phosphate.**— Shape of the crystals (knife-rest or coffin-lid), soluble in acids.

(*b*.) **Acid Ammonium Urate.**—Small dark balls, often covered with spines, and also amorphous granules (Fig. 32).

(c.) Cystin (very rare), hexa-gonal crystals, soluble in NH_4,HO (Fig. 31, a).

(c.) Carbonate of Lime.—Small colourless balls, often joined to each other; efferves-cence on adding acids (micro-scope).

(d.) Leucin and Tyrosin (very rare). (Fig. 32.)

(d.) Crystalline Phosphate of Lime.

(e.) Cholesterin (very rare). (Fig. 11.)

(e.) Leucin and Tyrosin (very rare).

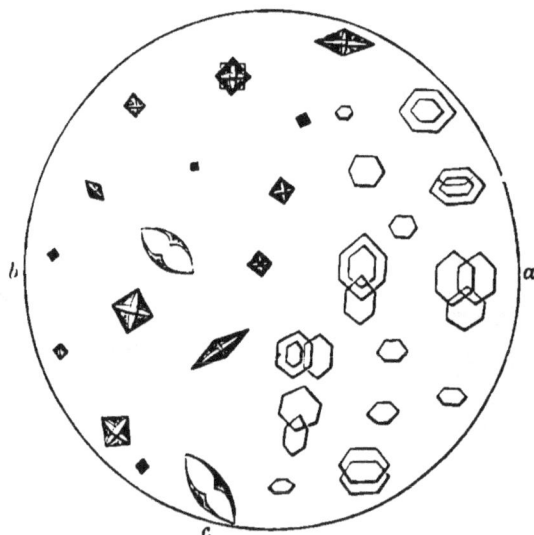

Fig. 31.—a, Crystals of cystin; b, oxalate of lime; c, hour-glass forms of b.

3. Urinary Calculi.

They are composed of urinary constituents which form urinary deposits, and may consist of one substance or of several, which are usually deposited in layers, in which case the most central part is spoken of as the "*nucleus.*" The nucleus not unfrequently consists of some colloid substance—mucus, a portion of blood-clot, or some albuminoid matter—in which crystals of oxalate of lime or globular urates become entangled. Layer after layer is then deposited. In certain cases the nucleus may consist of a foreign body introduced from without. Calculi are sometimes classified as *primary* and *secondary;* the former are due to some general alteration in the composition of the urine, whilst the latter are due to ammoniacal decomposition of the urine, resulting in the precipitation of phosphates on stones already formed. This of course has an important bearing on the treatment of calculous

Fig. 32.—*a a*, Leucin balls; *b b*, tyrosin sheaves; *c*, double balls of ammonium urate.

disorders. Calculi occur in acid and alkaline urine. A *highly acid* urine favours the formation of *uric acid* calculi, because that substance is most insoluble in very acid urine. A *highly alkaline* urine favours the formation of calculi, consisting of *calcium phosphate* or *triple phosphate*, as these substances are insoluble in alkaline urine.

4. Method of Examining a Calculus.

(*a.*) Make a section in order to see if it consists of one or more substances; examine it with the naked eye, and a portion microscopically.

(*b.*) Scrape off a little, and heat it to redness on platinum foil over a Bunsen burner.

(A) If it be *entirely combustible*, or almost so, it may

consist of uric acid or urate of ammonia, xanthin, cystin, coagulated fibrin or blood, or ureostealith.

(B) If *incombustible* or if it leaves *much ash*, it may consist of urates with a fixed base (Na, Mg, Ca), oxalate, carbonate, or phosphate of lime, or triple phosphate.

5. A. Combustible.—Of this group, uric acid and urate of ammonia give the *murexide test.*

(i.) **Uric Acid** is by far the most common form, and constitutes five-sixths of all renal concretions. Concretions the size of a split-pea, or smaller, may be discharged as *gravel.* When retained in the bladder they are usually spheroidal, elliptical, and somewhat flattened ; are tolerably hard ; the surface may be smooth or studded with fine tubercules ; the colour may be yellowish, reddish, or reddish-brown, very rarely white. When cut and polished, they usually exhibit a concentric arrangement of layers. Not unfrequently a uric acid calculus is covered with a layer of phosphates, and some calculi consist of alternate layers of uric acid and oxalate of lime. Its *chemical relations :* nearly insoluble in boiling water ; soluble in KHO, from which acetic acid precipitates uric acid crystals (microscopic); gives the murexide test (Lesson XVII., 3).

(ii.) **Urate of Ammonium Calculi** are very rare, and occur chiefly in the kidneys of children ; they form small irregular, soft, fawn-coloured masses, easily soluble in hot water.

(iii.) If the calculus is combustible and gives no murexide test it may consist of **Xanthin,** which is very rare, and of no practical importance.

(iv.) **Cystin** is very rare, has a smooth surface, dull yellow colour, which becomes greenish on exposure to the air ; and a glistening fracture with a peculiar soapy feeling to the fingers ; soft, and can be scratched with the nail. It occurs sometimes in several members of the same family. It is soluble in ammonia, and after evaporation it forms regular microscopic hexagonal plates (Fig. 31, *a*).

The other calculi of this group are very rare.

6. (A.) Group.—*Apply the Murexide Test.*

| It is obtained | Treat the original powder with potash. | No odour = *Uric Acid.*
Odour of HN_3 = *Ammonium Urate.* |

The residue is not coloured, but becomes yellowish-red on adding caustic potash . . . . } = *Xanthin.*

The residue is not coloured either by KHO or NH_4HO; the original substance is soluble in ammonia, and on evaporation yields hexagonal crystals . . . . } = *Cystin.*

On heating, it gives an odour of burned feathers; the substance is soluble in KHO, and is precipitated therefrom by excess of HNO_3 . . } = *Proteid.*

7. B. Incombustible.

(i.) Urates (Na, Ca, Mg), are rarely met with as the sole constituent. They give the murexide test.

(ii.) **Oxalate of Lime** or Mulberry Calculi, so called because their surface is usually tuberculated or warty ; they are hard, dark brown, or black. These calculi, from their shape, cause great irritation of the urinary mucous membrane. When in the form of gravel, the concretions are usually smooth, variable in size, pale grey in colour. Layers of oxalate of lime frequently alternate with uric acid. When heated it blackens, but does not fuse, and then becomes white, being converted into the carbonate and oxide. The white mass is alkaline to test-paper, and when treated with HCl, it effervesces (CO_2). Oxalate of lime is not dissolved by acetic acid.

(iii.) **Carbonate of Lime.**—Rare in man ; when met with they usually occur in large numbers. Dissolve with effervescence in HCl. Sometimes crystals occur as a deposit. They are common in the horse's urine.

(iv.) **Basic Phosphate of Lime Calculi** are very rare, and are white and chalky.

(v.) **Mixed Phosphates (Fusible Calculus)** consist of triple-phosphate and basic phosphate of lime. They indicate that the urine has been ammoniacal for some time, owing to decomposition of the urea. They are usually of considerable size, and whitish ; the consistence varies. When triple-phosphate is most abundant, they are soft and porous, but when the phosphate of lime is in excess, they are harder. A *whitish* deposit of phosphates is frequently found coating other calculi. This occurs when the urine becomes ammoniacal, hence in such cases regard must always be had to the condition of the urinary mucous membrane. Such calculi are incombustible, but, when exposed to a strong heat, fuse into a white enamel-like mass, hence the name, fusible calculi.

8. (B.) Group.

(i.) *The substance gives the murexide reaction, indicates urates.*

The residue is treated with water.

It is soluble, and the solution is alkaline, . .
- Neutralise; add platinic chloride, a yellow precipitate . . . . } = *Potash.*
- The residue yields a yellow flame . = *Sodium.*

Scarcely soluble; the solution is scarcely alkaline; soluble in acetic acid, .
- Ammonium oxalate gives a white crystalline precipitate . . . } = *Calcium.*
- Ammonium oxalate gives no precipitate, but on adding ammonium chloride, sodic phosphate, and ammonia there is a crystalline precipitate of triple-phosphate . . . } = *Magnesium.*

(ii.) *The original substance does not give the murexide test.*

Treat the original substance with hydrochloric acid.

It dissolves with effervescence . . . . = { *Calcic carbonate.* *Magnesic carbonate.*

It dissolves without effervescence. Heat the original substance, and treat it with HCl, . .
- It dissolves with effervescence . . = *Calcic oxalate.*
- There is no effervs'ce. Heat in a capsule .
  - It melts. The original stone treated with KHO. } Evolves NH₃ . } = *Triple phosphate.*
  - Evolves no NH₃ . } = *Neut. calc. phosp.*
  - It does not melt on heating . . = *Acid calc. phosp.*

9. General Examination of the Urine.

(i.) **Quantity** in twenty-four hours (normal 50 oz., or 1,500 cc.)

(ii.) **Colour, Odour, and Transparency** (if bile or blood be suspected, test for them).

(iii.) **Specific Gravity** of the mixed urine (if above 1030, test for sugar).

(iv.) **Reaction** (normally slightly acid ; if alkaline, is the alkali volatile or fixed ?).

(v.) **Heat.**

(*a.*) If a turbid urine becomes clear = *urates.*

(*b.*) If it becomes turbid = *earthy phosphates* or *albumin.* Albumin is precipitated before the boiling point is reached (70° C.), whilst phosphates are thrown down about the boiling point. It is necessary, however, to add HNO_3, which will dissolve the phosphates, but not the albumin. A case may occur where both urates and albumin are present ; on carefully heating, the urine will first become clear (urates), and then turbid, which turbidity will not disappear on adding HNO_3 (albumin). Estimate approximately the amount of albumin present.

(vi.) **Test for Chlorides,** with HNO_3 and $AgNO_3$ (if albumin be present, it must be removed by boiling and filtration).

(vii.) If sugar be suspected, test for sugar (Moore's, Trommer's, or Fehling's test), and if albumin be present, remove it.

(viii.) Make naked-eye, microscopic, and chemical examinations of the sediment.

APPENDIX.

Exercises on the foregoing.

A. The student must practise the **analysis of fluids** containing one or more of the substances referred to in the foregoing Lessons.

Suppose the solution contains one or more of the following—*Blood, bile, urea, uric acid,* or *ferments,* proceed as follows :—

(*a.*) If *blood* is suspected, use the spectroscope if the colour appears to indicate the presence of blood.

(*b.*) If the colour is such as to suggest the presence of *bile*, concentrate the fluid on a water-bath, and apply Gmelin's test for the bile-pigments (Lesson IX., 6). If proteids are absent, apply Pettenkofer's test for the bile-acids (Lesson IX., 3). If proteids are present, proceed as in Lesson I., p. 19.

(*c.*) In testing for *urea*, proceed as in Lesson XV., 2, by precipitating with baryta mixture. Filter, evaporate the filtrate to dryness, redissolve the residue with absolute alcohol, and allow some of the alcoholic extract to evaporate on a slide, and use the microscope for the detection of crystals of urea. Apply the other reactions for urea in Lesson XV., 4.

(*d.*) For *uric acid*, or its salts, add hydrochloric acid to precipitate the uric acid in crystals [Lesson XV. 2, (*a.*)], and to the latter apply the murexide test.

(*e.*) If *ferments* are suspected, their action must be tested on fibrin or starch mucilage, as the case may be, the reaction of the fluid being adapted to the ferment tested for.

B. If a powder or solid substance be given to you,

(*a.*) Examine it with the naked eye and microscopically, whether it be amorphous or crystalline.

(*b.*) Burn some in a tube; smell it to detect any odour. Observe if it leaves an ash.

(*c.*) Examine its solubility in water, caustic soda, salt solutions, alcohol, and ether.

(*d.*) Apply tests in Lesson I., 1, for *proteids*.

(*e.*) If it contain a proteid, test its solubility in water. *Native albumins*, *peptones*, and *gelatin* are soluble; the others are insoluble. Confirm by other tests in Lesson I.

(*f.*) If it be not a proteid, or if proteids be present, remove them. Test if it be *soluble in cold water*, and to the solution apply the tests for *dextrin* and *glycogen* (Lesson II., 4, 5), *reducing sugars, e.g.*, grape or milk sugars (Lesson II., 6). The former, in a concentrated solution, is precipitated by absolute alcohol, the latter is not. If cane-sugar be suspected, invert it (Lesson II., 9, *d.*), and test for a reducing sugar. Test also for *urea* (Lesson XV., 4, 7).

(*g.*) Ascertain its solubility in warm water, *starch* (Lesson II., 1), *urates* (Lesson XVII., 5), *tyrosin* (Lesson VIII., 5).

(*h.*) Test for uric acid (Lesson XVII., 3).

(*i.*) *Cholesterin* is insoluble in cold water and alcohol, but soluble in ether. On evaporation of the ether, the characteristic crystals are obtained (Lesson IX., 7).

(*j.*) Fats melt on heating and are soluble in ether, leaving a greasy stain (Lesson IV., 13).

C. Analysis of Urine.—The student must also practise the analysis of urines containing one or more abnormal constituents, and he must also practise the estimation of the quantity of the more important substances present. Both sets of processes must be done over and over again, in order that he may perfect himself in the methods in common use.

PART II.—EXPERIMENTAL PHYSIOLOGY.

Before beginning the experimental part of the course, each student must provide himself with the following :—A large and a small pair of scissors ; a large and a fine pointed pair of forceps ; a scalpel ; a blunt needle or "seeker" in a handle ; pins ; fine silk thread ; bees'-wax ; sealing wax ; two camel's-hair brushes of medium size. It is convenient to have them all arranged in a small case.

PHYSIOLOGY OF MUSCLE AND NERVE.

LESSON XXII.

GALVANIC BATTERIES AND GALVANOSCOPE.

1. **Daniell's Cell** consists of a glazed earthenware pot with a handle (Fig. 33) and containing a saturated solution of cupric sulphate. Some crystals of cupric sulphate are placed in it to keep the solution saturated. The pot itself is about 18 cm. high, and 9 cm. in diameter. In the copper solution is placed a roll of sheet-copper, provided with a tongue, to which a binding screw is attached. Within is a porous unglazed cylindrical cell containing 10 per cent. solution of sulphuric acid. A well amalgamated rod of zinc provided at its free end with a binding screw is immersed in the acid. The zinc is the negative (−), and the copper the positive (+) pole.

Fig. 33.—Daniell's Cell.

2. **Amalgamation of the Zinc.**—The zinc should always be well amalgamated. When a cell hisses, the zinc requires

to be amalgamated. Dip the zinc in 10 per cent. sulphuric acid until effervescence commences. Lift it out and place it on a shallow porcelain plate. Pour some mercury on the zinc, and with a piece of cloth rub the mercury well over the zinc. Dip the zinc in the acid again, and then scrub the surface with a rag under a stream of water from the tap. Collect all the surplus mercury and place it in the bottle labelled "Amalgamation Mixture." Take care that none of the mercury gets into the soil pipe. A very convenient method is to dip the zinc into a glass tube with thick walls containing mercury and sulphuric acid. For convenience the tube is fixed in a block of wood.

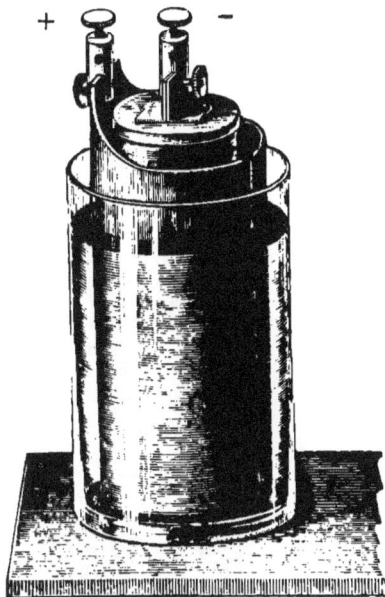

3. Grove's Cell· (Fig 34) consists of an outer glazed earthenware glass or ebonite vessel containing a roll of amalgamated zinc and dilute (10 per cent.) sulphuric acid.

Fig. 34.—Large Grove's Element.

In the inner porous cell is placed platinum foil with strong nitric acid. The platinum and zinc are provided with binding screws. The platinum is the + positive pole or anode, the zinc the − negative pole or cathode.

For physiological purposes, the small Grove's cells about 7 cm. in diameter and 5 cm. in height, are very convenient. When in use the battery ought to be placed in a draught chamber to prevent the nitrous fumes from affecting the experimenter.

4. Bichromate Cell (Fig. 35).—This consists of a glass bottle containing one zinc and two carbon plates immersed in the following mixture:—Dissolve 1 part of potassic bichromate in 8 parts of water and add 1 part of sulphuric acid. The zinc is attached to a rod, which can be raised when it is desired to stop the action of the battery. This cell is convenient enough when it is not necessary to use a current of perfectly constant intensity.

Other forms of batteries are used, but the foregoing are sufficient for the purposes of these exercises.

5. The Galvanoscope or Detector.

(*a.*) Charge a Daniell's cell and attach a copper wire to the negative pole (zinc), and another to the positive pole (copper). On bringing the free ends of the two wires together, the circuit is made, and a current of continuous, galvanic, or voltaic electricity circulates outside the battery from the + to the − pole. The existence of this electrical current may be proved in many ways—*e.g.*, by the effect of the current on a magnetic needle.

(*b.*) Use a vertical gal-

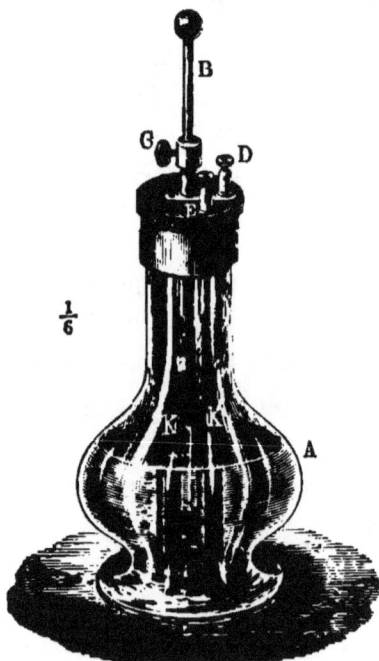

Fig. 35.—Bichromate Cell.—A, the glass vessel; K, K, carbon; Z, zinc; D, E, binding screws for the wires; B, rod to raise or depress the zinc in the fluid; C, screw to fix B.

Fig. 36.—Detector.

vanoscope, or as it is called by telegraphists, a **detector** (Fig. 36), in which the magnetic needle is so loaded as to rest in a vertical position. A needle attached to this moves over a semi-circle graduated into degrees. Connect the wires from the + and − poles of the Daniell's battery with the binding screws of this instrument, and note that when the circuit is made, the needle is deflected from its vertical into a more or less horizontal position, but the angle of deflection is not directly proportional to the current passing in the instrument. Break the circuit by removing one wire, and notice that the needle travels to zero and resumes its vertical position. The detector made by Stöhrer, of Leipzig, is a convenient form.

LESSON XXIII.

ELECTRICAL KEYS, RHEOCHORD.

It is convenient to make or break—*i.e.*, close or open—a current by means of keys of which there are various forms.

1. Du Bois Key (Fig. 37).—It consists of a square plate of vulcanite, attached to a wooden or metallic framework which can be screwed to a table. Two oblong brass bars (II. and III.), each provided with two binding screws, are fixed to the ebonite, while a movable brass bar (IV.) with an ebonite handle is fixed to one of the bars, and can be depressed so as to touch the other brass bar.

Two Ways of Using the du Bois Key.

2. (1) *When the key is closed the*

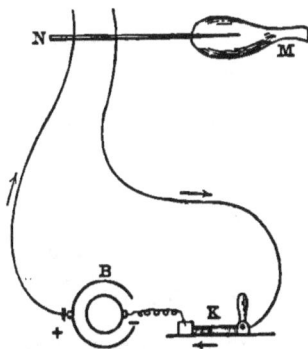

Fig. 37.—Du Bois - Reymond's Friction Key.

Fig. 38.—Scheme of du Bois Key.—B, battery; K, key; N, nerve; M, muscle.

current is made, and when it is opened the current is broken
(Fig. 38).   **Apparatus.**—Use a charged Daniell's cell and detector
as before, three wires, and a du Bois key screwed to a table.

(*a.*) As in the scheme (Fig. 38) connect one wire from
the battery to one brass bar of the key.   Connect the other
brass bar with one binding screw of the detector.   Connect
by means of the third wire the other binding screw of the
detector with the − pole of the cell.

(*b.*) Observe on depressing the key (*i.e.*, making the
circuit) the needle is deflected, on raising it (*i.e.*, breaking
the circuit) the needle passes to zero.   This method of using
the key we may call that for "making and breaking a
current."

This method is never used for an interrupted current applied
to a nerve or muscle.

3. (2) *When the key is closed the current is said to be* "**short-
circuited.**"   **Apparatus.**—Daniell's cell, detector, four wires, and
a du Bois key screwed to the table.

(*a.*) As in the scheme (Fig. 39) connect the wire from
the positive pole of the battery
to the outer binding screw of one
brass bar of the key, and the
other battery wire to the outer
binding screw of the other brass
bar of the key.   Then connect
the inner binding screws of both
brass bars with the binding
screws of the detector.

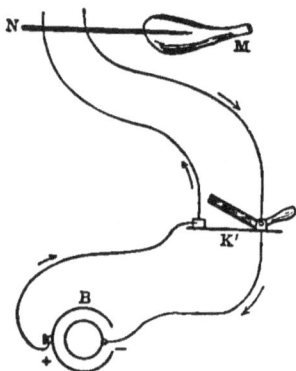

Fig. 39.—Scheme of du Bois
Key for Short-circuiting.—
N, nerve ; M, muscle ; B,
battery ; K', key.

(*b.*) Observe when the key is
depressed or closed, there is no
deflection of the needle—*i.e.*,
when the current is cut off from
the circuit beyond the key or
bridge; when the key is raised,
the needle is deflected.

When the key is depressed,
the current is said to be "**short-circuited,**" for the key acts like a
bridge, and so a large part of the current passes through it back
to the battery, while only an excessively feeble current passes

through the wires beyond the key, so feeble is it that it does not affect a nerve. On raising the key, the whole of the current passes through the detector or nerve, as the case may be. This method of using the key we may call the method of "short-circuiting."

N.B.—In using the key to apply an induction current to excite a nerve or muscle, always use this key by the second method—*i.e.*, always short-circuit an induction current.

4. Mercurial Key.—Where a *fluid* contact is required the wires dip into mercury. Study the use of this key. It is used in the same way as a du Bois key.

5. Plug Key (Fig. 40).—Two brass bars are fixed to a piece of vulcanite. The circuit is made or broken by inserting a brass plug between the bars. Each brass bar is provided with two binding screws, to which one or two wires may be attached, so that it can be used like a du Bois key, either by the first or second method.

Fig. 40.—Plug Key.

6. Morse Key (Fig. 41).—If it is desired to make or break a current rapidly this key is very convenient. If this key be used to make and break the primary circuit, connect the wires to B and C; when the style of the lever, *l*, is in contact with *c*, the current does not pass in the primary circuit. On depressing the handle, K, the primary circuit is made. If, however, the wires be

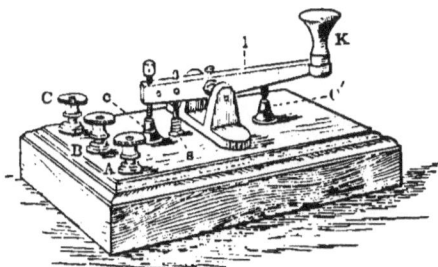

Fig. 41.—The Morse Key.—The connections are concealed below, but are B to *l*, A to *c*, C to *c'*.

connected to A and B the current passes and is broken on depressing K. To use this key as a short-circuiting key, connect the wires from the battery to A and B, and those of the electrodes to B and C. The current is short-circuited until K is depressed, when the current passes from C to B through the electrode wires.

7. The "Trigger or Turn-over Key" is referred to in Lesson **XXXII.**

8. The **Contact** or **Spring Key** (Fig. 42) is also very useful for rapidly making and breaking a circuit. The current can only pass between the binding screws when the metallic spring is pressed down. The left end of the spring is in metallic contact with the upper binding screw, while the second binding screw is similarly connected with the little metallic peg at the right-hand end of the fig.

Fig. 42.—Spring Key.

9. **Means of Graduating a Galvanic Current.**—Besides altering the number, arrangement, or size of the cells themselves, we can use an arrangement to divide the current itself, the battery remaining constant. This is effected by the simple rheocord.

The **Simple Rheochord** consists of a brass or German-silver wire about 1 metre in length, stretched longitudinally along a board, and with its ends connected to binding screws and insulated (Fig. 43). On the wire there is a "slider" which can be pushed along as desired. **Apparatus.** —Simple rheochord, Daniell's cell, detector, du Bois key, five wires.

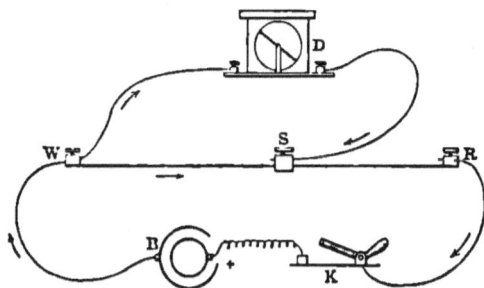

Fig. 43.—Scheme of Simple Rheochord.—B, Battery ; K, key ; W, R, wire ; S, slider ; D, detector.

(*a.*) Arrange the experiment as in Fig. 43. When the slider S is hard up to W, practically all the electricity passes along the wire (W, R), back to the battery.

(*b.*) Pull the slider away from W, and in doing so, more resistance is thrown into the battery circuit, and some of the electricity passes along the detector circuit and deflects the needle. The deflection is greater—but not proportionally so—the further the slider is moved from W.

10. Use in the same way a rheochord with two platinum wires, which are connected by an ebonite cup filled with mercury, which

slides along on the two wires. Connect the battery—a key being interposed in one wire—with the binding screws on one end of the rheocord, and to the same binding screws connect two wires to the detector. Observe as the mercury cup is pulled away from the binding screws, there is a greater deflection of the needle, but the deflection is not in proportion to the distance of the cup.

11. The **Rheochord** of du Bois-Reymond is used to vary the amount of a *constant current* applied to a muscle or nerve. It consists of a long board or box, with German-silver wire—of varying length and whose resistance is accurately graduated—stretched upon it. At one end are a series of brass blocks disconnected with each other above, but connected below by a German-silver wire passing round a pin. These blocks, however, may be connected directly by brass plugs $S_1 S_2 \ldots S_5$. From the blocks 1 and 2, two platinum wires pass from A to the opposite end of the box (Y), where they are insulated. Between the wires is a " slider " (L), consisting of two cups of mercury, which slide along the the wires.

In using the instrument, take a Daniell's battery and connect its wires to the binding screws at A and B, and to the same screws attach the wires of the electrodes over which the nerve (c d) of the muscle (F) is laid. We have two circuits (a c d b and a A B b), the wires of the rheochord introduced into the latter.

Push up the slider with its cups (L) until it touches the two brass plates 1 and 2, and insert all the plugs (S_1–S_5) in their places, thus making the several blocks of brass practically one block. In this position the resistance offered by the rheochord circuit is so small as compared with that including the nerve, that practically all the

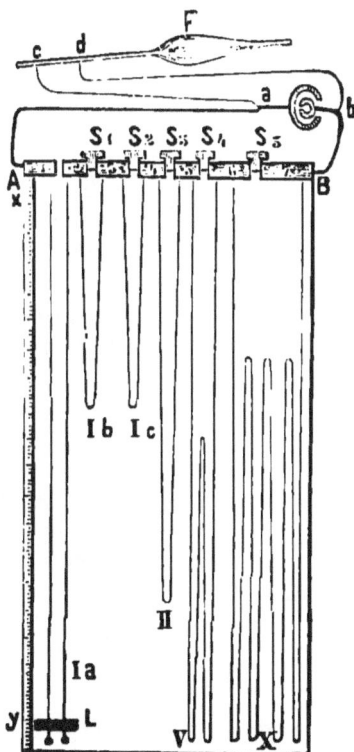

Fig 44.—Rheochord of du Bois-Reymond.

electricity passes through the former and none through the latter.

Move the slider away from A, when a certain resistance is thrown into the rheochord circuit, according to the length of the platinum wires thus introduced into it, and so a certain fraction of the current is sent through the electrode circuit. If the plug S_1 be taken out more resistance is introduced, that due to the German-silver wire (I b), and, therefore, a certain amount of the current is made to pass through the electrode circuit. By taking out plug after plug more and more resistance is thrown into the rheocord circuit. The plugs are numbered, and the diameter and length of the German-silver wires are so selected in making the instrument, that the resistances represented by the several plugs when removed, are all multiples of the resistance in the platinum wires on which the slider moves. The instrument is described here for convenience, but its use will be practised later.

(a). Connect a Daniell's cell or a small Grove with the binding screws at A and B, introducing a du Bois key in the circuit. To A and B attach two other wires, and connect them to a du Bois key, and to the key attach electrodes, thus short-circuiting the electrode circuit.

(b). Prepare a nerve muscle preparation, lay the muscle on glass, and place the nerve over the electrodes.

(c), Put in all the plugs and push up the slider close to the blocks. Open the short-circuiting key. Make the battery circuit, perhaps a contraction may be obtained. Pull the slider away from the blocks, and on making the current contraction will occur, and perhaps also on breaking it. Take out plug S_1, and pull the bridge still further away, and very probably there will be contraction both at make and break. Proceed taking out plug after plug, and note the result. The result, and explanation thereof will be referred to in Lesson XLII., 2.

12. Pohl's Commutator.—Sometimes it is desired to send a current through either of two pairs of wires. This is done by means of Pohl's commutator without the cross-bars (Lesson XXX.,

Fig. 59). At other times it is desired to reverse the direction of a current. This is done by Pohl's commutator with cross-bars, or by means of Thomson's reverser.

13. Thomson's Reverser (Fig. 45) may be used to reverse the direction of a constant current. The wires from the battery are connected to the two lower, and those from the electrodes to the upper binding screws. The binding screws are four in number, and placed behind the circular disc seen in the figure. When the handle is horizontal the current is shut off from the electrodes, while the direction of the current is reversed by raising or lowering the handle. This instrument is used solely for reversing the direction of a current.

Fig 45.
Thomson's Reverser.

LESSON XXIV.

INDUCTION MACHINE—ELECTRODES.

1. **Induced or Faradic Electricity** is most frequently employed for physiological purposes.

2. **Induction Apparatus of du Bois-Reymond.**—In Fig. 46 the primary coil (R') consists of about 150 coils of thick insulated copper wire, the wire being thick to offer slight resistance to the galvanic current. The secondary coil (R″) consists of 6000 turns of thin insulated copper wire arranged on a wooden bobbin ; the whole spiral can be moved along the board (B) to which a millimetre scale (I) is attached, so that the distance of the secondary from the primary spiral may be ascertained. At the left end of apparatus is Wagner's hammer as adapted by Neef, which is an automatic arrangement for opening and breaking the primary circuit. When Neef's hammer is used to obtain what is called an interrupted current, the wires from the battery are connected as in the figure, but when single shocks are required, the wires

from the battery are connected with a key, and this again with the two terminals of the primary spiral, S″ and S‴.

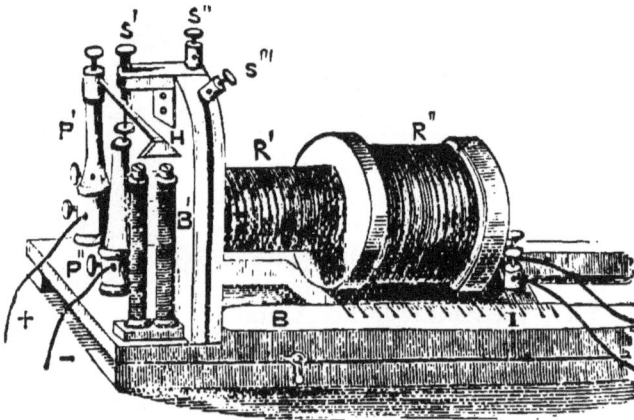

Fig. 46.—Induction apparatus of du Bois-Reymond. — R′, Primary; R″, secondary spiral; B, board on which R″ moves; I, scale; + -, wires from battery; P′, P″, pillars; H, Neef's hammer; B′, electro-magnet; S′, binding screw touching the steel spring (H); S″ and S‴, binding screws to which are attached wires when Neef's hammer is not required.

3. **Ordinary Hand Electrodes.**—Take two pieces of flexible gutta-percha coated wire (No. 20) 60 cm. long, and two pieces of thick glass tubing 8 cm. long, and with a bore sufficient to admit the wire. Push a wire through each tube, and allow the end of the wire to project 3 cm. beyond the tube, scrape the gutta-percha off the free ends of both wires. Fix the wires in the glass tube with sealing-wax, and with a well-waxed thread bind the two tubes together. A very handy holder is made by thrusting two fine insulated wires (No. 36) through the bone handle of a crotchet needle.

4. For some purposes " **shielded electrodes** " are used, i.e., the platinum terminals are exposed only on one side, the other being sunk in a piece of vulcanite (Fig. 61, E).

5. **Du Bois-Reymond's Electrodes** (Fig. 47).—The two wires end in triangular pieces of platinum (P), which rest on a glass plate. The whole is supported on a stand (V), and can be moved in any direction by the universal joint (B).

6. For Non-polarisable Electrodes see Lesson XXXVIII., 2.

7. Polarisation of Electrodes—Apparatus.—Pair of ordinary electrodes, two wires, du Bois key, spring key, Daniell's cell, frog, and instruments.

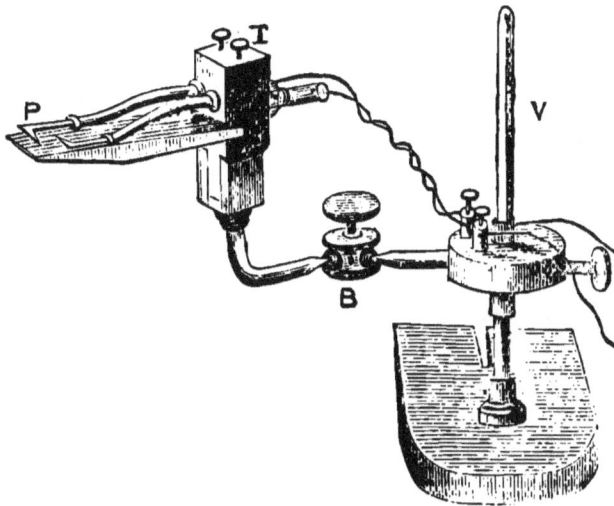

Fig. 47.—Du Bois-Reymond's Platinum Electrodes. The nerve is placed over the two pieces of platinum, P, which rest on glass ; B, universal joint; V, support.

(*a.*) Pith a frog (Lesson XX., 1), lay it belly downwards on a frog plate, and expose one sciatic nerve.

(*b.*) Clamp the du Bois key to the table, place the electrodes under the sciatic nerve, and connect their other ends each with the outer binding screw of the brass plates in the du Bois key. Close the key, and observe that no contraction of the leg muscles occurs.

(*c.*) By two wires connect a Daniell's cell with the du Bois key, introducing a spring key in the circuit. Open the key to allow the constant current to pass through the nerve for three minutes or thereby, and observe that there is no contraction as long as the constant current is passing. Close the key, *i.e.*, short-circuit the battery, and at once a contraction occurs. Remove the battery, close and open the key.

Contractions occur, but they gradually get feebler. The contractions are due to polarisation of the electrodes.

LESSON XXV.

SINGLE INDUCTION SHOCKS—INTER-RUPTED CURRENT—BREAK EXTRA-CURRENT—HELMHOLTZ'S MODIFICATION.

1. **Single Induction Shocks—Apparatus.**—Grove's cell charged, induction machine, five wires, two du Bois keys, and ordinary electrodes.

(*a.*) Connect one wire from the battery to the binding screw, *s''*, of the induction machine (Fig. 46). Join the other wire from the battery to a du Bois key, and use the third wire to connect the key with the binding screw, *s'''*. The key is used according to the first method, *i.e.*, for make and break, so that the primary current can be made or broken at will. To the binding screws of the secondary coil attach two wires, and connect them to the short-circuiting du Bois key, and to the latter the electrodes as in Fig. 54.

(*b.*) Open the short-circuiting key, push the secondary coil pretty near to the primary, and place the points of the electrodes on the tip of the tongue, or hold them between the forefinger and thumb moistened with water. Close the key in the primary circuit, *i.e.*, make the circuit, and instantaneously at the moment of making, a shock is induced in the secondary coil, R'', and is felt on the tip of the tongue or finger. It is called the closing or **make induction shock.** All the time the key is closed the galvanic current is circulating in the primary spiral, but it is only when the primary current is made or broken that a shock is induced in the secondary spiral.

(*c.*) **Break the primary** current by raising the key, and

THE BREAK EXTRA-CURRENT OF FARADAY.

instantaneously a shock is felt as before ; this is the opening or **break induction shock.**

(*d.*) Observe that the *break is stronger than the make shock.* Push the secondary coil a long distance from the primary, and with the electrodes on the tongue, make and break the primary circuit, and gradually move the secondary near the primary coil. Observe that the break shock is felt first, and on pushing the secondary nearer the primary coil both shocks are felt, but the break is stronger than the make shock.

2. Interrupted Current by using Neef's Hammer—Apparatus.— The same as for single shocks.

(*a.*) Connect the battery wires as in Fig. 46, *i.e.*, to the binding screws in the erect pillars, P′ (+) and P″ (-). Introduce a du Bois key as for the make and break arrangement. The automatic vibrating spring, or Neef's hammer, is now included in the primary circuit. Set the spring vibrating. Make the current by depressing the handle of the key. The elastic spring, H, is attracted by the temporary magnet, B′, thus breaking the contact between the spring, H, and the screw, S′, and causing a break shock in the secondary coil. B′ is instantly demagnetised, the elastic spring recoils and makes connection with S′, and causes a make shock. Thus a series of make and break induction shocks following each other with great rapidity is obtained, but the make and break shocks are in alternately opposite directions.

(*b.*) While Neef's hammer is vibrating, apply the electrodes to the tongue or finger as before, noting the effect produced and how it varies on altering the distance between the secondary and primary spirals.

3. The Break Extra-Current of Faraday.—When a galvanic current traversing the primary coil of an induction machine is made or broken, each turn of the wire exerts an inductive influence on the others. When the current is *made*, the direction of the extra current is *against* that of the battery current, but at *break* it is in the same direction as-the battery current.—**Apparatus.**—Daniell's cell, two du Bois keys, five wires, induction coil, ordinary electrodes (and nerve muscle preparation).

(*a.*) Arrange the apparatus according to the scheme

(Fig. 48). Notice that both keys and the primary coil of the induction machine are in the primary circuit, both keys being so arranged that either the primary coil, P, or the electrodes attached to key K', can be short-circuited.

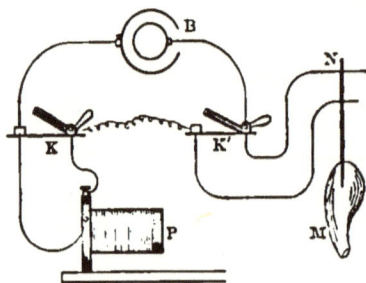

Fig. 48.—Scheme of the Break Extra-current.—B. Battery; K and K', keys; P, primary coil; N, nerve.

(b.) Test (α) either by electrodes applied to the tongue, or (β) by means of a nerve muscle preparation (β to be done after the student has learned how to make a nerve-muscle preparation).

(c.) Close the key, K, thus short-circuiting the coil. Open and close key, K'. There is very little effect.

(d.) Open K', the current passes continuously through the primary coil. Open key K, a marked sensation is felt, due to the break extra-current.

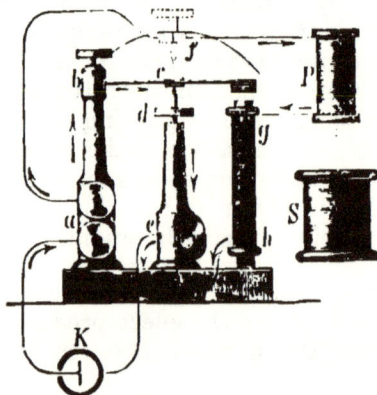

Fig. 49.— Helmholtz's Modification of Neef's Hammer.—As long as c is not in contact with d, g h remains magnetic; thus c is attracted to d, and a secondary circuit, a, b, c, d, e is formed; c then springs back again, aud thus the process goes on. A new wire is introduced to connect a with ƒ. K, battery.

4. Helmholtz's Modification.

—The break shock is stronger than the make, and to equalise them Helmholtz devised the following modification :—

(a.) Connect the battery wires as before to the two pillars (Fig. 46), P' and P", or to a and e (Fig. 49). In Fig. 49 connect a wire from a to ƒ, thus bridging or "short-circuiting" the interrupter. Elevate the screw (ƒ) out of reach of the spring (c), but raise the screw (d) until it touches the spring at every vibration. By this means the make and break shocks are nearly equalised, but both shocks are weaker.

LESSON XXVI.

PITHING—CILIARY MOTION—NERVE-MUSCLE PREPARATION—NORMAL SALINE.

1. To Pith a Frog.—Wrap the body, fore and hind legs, in a towel, leaving the head projecting. Grasp the towel enclosing the frog with the little, ring, and middle fingers and thumb of the left hand, leaving the index finger free. With the index finger bend down the frog's head over the inner surface of the second finger until the skin over the back of the neck is put on the stretch. With the nail of the right index finger feel for a depression where the occiput joins the atlas, marking the position of the occipito-atlantoid membrane. With a sharp, narrow knife held in the right hand, divide the skin, membrane, and the medulla oblongata. Withdraw the knife, thrust a "seeker" into the brain cavity through the opening just made, and destroy the brain. To prevent oozing of blood, a piece of a wooden match may be thrust into the brain cavity. If it is desired, destroy also the spinal cord with the seeker or a wire. The knife used must not have too broad a blade, else two large blood-vessels will be injured. The operation should be performed without losing any blood.

2. Ciliary Motion.

(*a.*) Pith a frog, destroying the brain and the spinal cord. Place the frog on its back. Divide the lower jaw longitudinally, and carry the incision backwards through the pharynx and œsophagus. Pin back the flaps. Moisten the mucous membrane, if necessary, with normal saline.

(*b.*) Make a small cork flag, and rest it on the mucous membrane covering the hard palate between the eyes. It will be rapidly carried backwards by ciliary motion towards the stomach. Repeat the experiment, and determine the time the flag takes to travel a given distance.

(*c.*) Grains of charcoal or Berlin blue are carried backwards in a similar manner.

10

(*d.*) Apply heat to the preparation, and observe that the cork travels much faster.

3. Anatomy of the Nerve-Muscle Preparation.—Before making

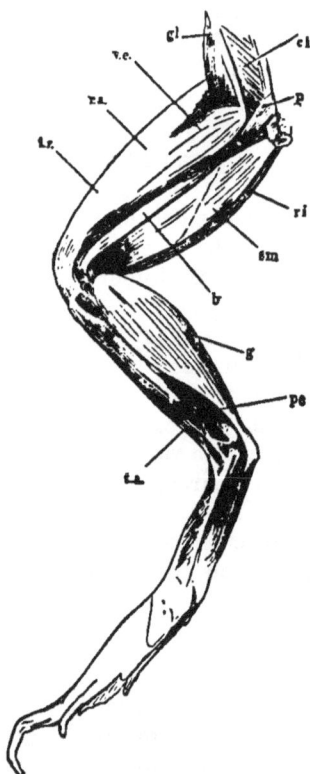

Fig. 50.—The Muscles of the left leg of a frog from behind.—*c.i.*, Coccygeo-iliacus; *gl*, gluteus; *p*, pyriformis; *r.a.*, rectus anterior; *v.e.*, vastus externus; *tr*, triceps; *r.i.″*, rect. int. minor; *s.m.*, semi-membranosus; *b*, biceps; *g*, gastrocnemius; *t.a.*, tibialis anticus; *pe*, peroneus.

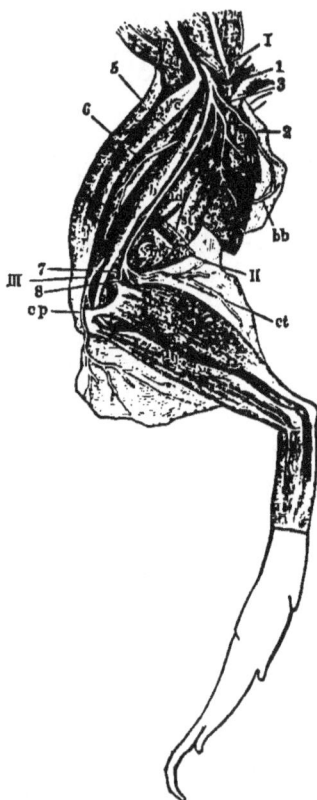

Fig. 51.—Distribution of the Sciatic Nerve (I.) of the Frog (see also Fig. 50).—*s.t.*, Semi-tendinosus; *ad″*, adductor magnus; (II.) its tibial; and (III.) peroneal divisions. ·

this preparation the student must familiarise himself with the anatomy of the lower limb of the frog. On a dead frog, or on a dissection of a frog already prepared, study the arrangement of the muscles, as shown in Fig. 50. The skin of the frog is supposed to be removed, and the frog placed on its belly, and the muscles viewed from behind. On the outside of the thigh, the *triceps femoris* (*t.r*), composed of the *rectus anterior* (*r.a*), the *vastus externus* (*v.e*), and the *vastus internus*, not seen from behind. On the median side, the *semi-membranosus* (*s.m*), and between the two the small narrow *biceps* (*b*). Notice, also, the *coccygeo-iliacus* (*c.i*), the *gluteus* (*gl*), the *pyriformis* (*p*), and the *rectus internus minor* (*r.i*).

In the leg, the *gastrocnemius* (*g*), with its tendo achillis, the *tibialis anticus* (*t.a*), and the *peroneus* (*pe*).

4. Make a Dissection.

(*a.*) Remove the skin from the leg of a dead frog; with a blunt needle, called a "seeker" or a "finder," gently tear through the fascia covering the thigh muscles, and with the blunt point of the finder separate the semi-membranosus from the biceps, and in the interval between them observe the *sciatic nerve* and the *femoral vessels*. Carefully isolate both, beginning at the knee, where the nerve divides into two branches—the tibial and peroneal—and work upwards (Fig. 51).

(*b.*) Follow the nerve right upwards to its connection with the vertebral column, and observe that it is necessary to divide the pyriformis (*p*) with small scissors, and also the ilio-coccygeal muscle, when the three spinal nerves—the 7th, 8th, and 9th—which form the sciatic nerve, come into view.

5. **Normal Saline.**—Dissolve 7·5 grammes of dried sodic chloride in 1000 cc. of distilled water. This is the best fluid to use to moisten tissues.

LESSON XXVII.

NERVE-MUSCLE PREPARATION—STIMU-
LATION OF NERVE—MECHANICAL,
CHEMICAL, AND THERMAL
STIMULI.

1. **Prepare a Nerve-Muscle Preparation—Apparatus.**—A frog, pithing-needle, or "seeker" in handle, narrow-bladed scalpel, a small and a large pair of scissors, forceps, towel, and a porcelain plate on which to place the frog.

(*a.*) Pith a frog with a pithing-needle, destroying the brain and spinal cord, and place the frog on its belly on a porcelain plate. With the small scissors make an incision through the skin along the back of one thigh—say the left—from the knee to the lower end of the coccyx, and prolong the incision along the back, a little to the left of the urostyle. Reflect the skin, which is readily done, as there are large lymph sacs between it and the muscles, thus exposing the muscles shown in Fig. 50.

(*b.*) Gently separate the semi-membranosus and biceps with the "seeker," and bring into view the sciatic nerve and femoral vessels. Still working with the seeker, and beginning near the knee, clear the sciatic nerve of any connective tissue around, but on no account is the nerve to be touched with forceps, nor is it to be scratched or stretched. With the small pair of scissors divide the pyriformis and ilio-coccygeus, and trace the nerve up to the vertebral column.

(*c.*) With a sharp-pointed, large pair of scissors divide the spinal column above the seventh lumbar vertebra; seize the tip of the urostyle with forceps, raise it, and with the strong scissors cut it clear from all its connections as far as the last lumbar vertebra, and then divide the urostyle itself. Divide the left iliac bone above and below, and remove it with the muscles attached to it. The lumbar plexus now comes

clearly into view. Bisect lengthways the three lower vertebræ, and use the quadrilateral piece of bone by which to
manipulate the nerve. With the forceps lift the fragment
of bone, and with it the sciatic nerve ; trace the latter downwards to the knee, dividing any connections and branches
with fine scissors. If the parts tend to become dry, moisten
the whole preparation with normal saline solution.

(d.) Divide the skin over the gastrocnemius, and expose this muscle.
Divide the tendo achillis below its
fibro-cartilage, lift the tendon with a
pair of forceps, and detach the gastrocnemius from its connections as far up
as the lower end of the femur. Cut
across the knee-joint, and remove the
tibia and fibula with their attached
muscles. Taking care to preserve the
sciatic nerve from injury, clear the
muscles away from the lower end of
the femur, and then divide, with the
large pair of scissors, the femur itself
about its middle. This preparation
(Fig. 52) consists, therefore, of the
gastrocnemius, and the whole length
of the sciatic nerve, to which is attached a fragment of bone, by which
the preparation can be manipulated
without injuring the nerve.

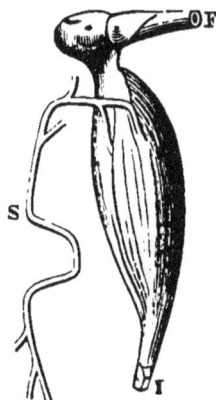

Fig. 52.—Nerve-muscle
Preparation.—S, Sci-
atic nerve—the frag-
ment of the spinal
column is not shown ;
F, femur; and I, tendo
achillis.

N.B.—The nerve must not be touched with instruments,
neither stretched nor scratched, nor allowed to come into contact
with the skin, and it must be kept moist with normal saline.

(a.) **Another method** is sometimes adopted. Pith a frog.
With the left hand seize the hind limbs and hold the frog
with its belly downwards. With one blade of a sharp-
pointed pair of scissors transfix the body immediately
behind the shoulder-blades, and divide the spinal column.
The head now hangs down, and by its weight it pulls the
ventral from the dorsal parts.

(b.) With the scissors divide the wall of the abdomen on
both sides parallel to the vertebral column, and remove all

the abdominal viscera. With the left hand seize the upper end of the divided spinal column, and with the right the skin covering it, and pull. At once the whole lower end of the trunk and the lower limbs are denuded of skin.

(c.) Take the thigh muscles between the thumb and fore-finger of the left hand, and with the point of one blade of a pair of scissors tear through the fascia between the biceps and semi-membranosus to expose the sciatic nerve, and then proceed as directed in 1.

2. **Stimuli** may be classified as follows :—

(1.) *Mechanical—e.g.*, cutting or pinching a nerve or muscle.

(2.) *Chemical—e.g.*, by dipping the end of a nerve in a saturated solution of common salt.

(3.) *Thermal—e.g.*, applying the end of a heated wire to the nerve.

(4.) *Electrical—*
    (a.) Continuous current.
    (b.) Single induction shocks.
    (c.) Interrupted current.

3. **Stimulation of Muscle and Nerve.**—It is convenient to modify somewhat the physiological limb, in order to render the muscular contraction more visible.—**Apparatus.**—Frog-pithing needle, scalpel, scissors, forceps, straw-flag, pins, muscle-forceps, camel's-hair brush, saturated solution of common salt in a glass thimble, ammonia, copper wire, spirit lamp or gas flame.

4. **Mechanical Stimulation.**

(a.) Pith a frog, destroying its brain and spinal cord. Remove the skin, and proceed as directed by the method (Lesson XXVII., 1) for preparing a nerve-muscle prepara-tion, as far as the isolation of the sciatic nerve, but modify the subsequent details as follows :—

(b.) After the nerve is cleared as far as the spine, clear all the muscles away from the femur, and divide the latter about its middle. Divide the sciatic nerve as high up as possible. The preparation consists of the leg, the lower end of the femur, and the sciatic nerve, terminating in the leg muscles. Pin a straw-flag to the toes by means of two pins. Fix the femur in a clamp or pair of muscle-forceps supported on a stand, and shown in Fig. 53, taking care that the gastrocnemius is upwards. The nerve hangs down, as shown in the figure, and must always be manipulated with a camel's-hair brush dipped in normal saline.

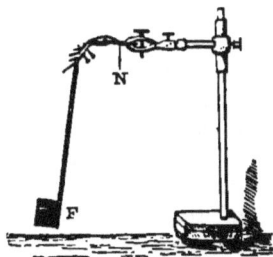

Fig. 53.—Straw-flag attached to a frog's leg fixed in a clamp.—N, Nerve; F, flag.

(c.) Pinch the free end of the nerve sharply with forceps, the muscles contract and the straw-flag is suddenly raised. Cut off the killed part of the nerve, and observe that contraction also occurs.

(d.) Prick the muscle with a needle, it contracts.

5. Thermal Stimulation.

(a.) To the same preparation apply gently, either to muscle or nerve, a hot copper wire or needle heated to a dull heat, a contraction results in either case. Cut off the dead part of the nerve.

6. Chemical Stimulation.

(a.) Place some saturated solution of common salt in a small glass thimble, or place a drop on a perfectly clean glass slide, and allow the free end of the nerve to dip into it. Owing to the high specific gravity of the saline solution, the nerve floats on the surface, but sufficient salt diffuses into the nerve to stimulate it. After a few moments, the individual joints of the toes begin to twitch, and by-and-by the whole limb is thrown into irregular spasms, ultimately terminating in a powerful, more or less continuous, contraction or spasm of the whole musculature, constituting *tetanus.* Cut off the part of the nerve affected by the salt, and the spasms will cease.

(*b.*) Use the same preparation, cover the leg with the skin of the frog, or wrap it in blotting-paper saturated with normal saline. Expose the fresh cut end of the nerve to the vapour of strong ammonia; there is no contraction of the muscle, but the ammonia kills the nerve. Instead of doing this, the whole leg may be laid on a card, covered with blotting-paper moistened with normal saline, with a hole in it just sufficient to allow the sciatic nerve to pass through it. The card is placed over a test-tube containing a drop of ammonia; the nerve hanging in the vapour of the latter is speedily killed, but there is no contraction of the muscle. Apply the ammonia to the muscle, it will contract.

LESSON XXVIII.

SINGLE AND INTERRUPTED INDUCTION SHOCKS—TETANUS—CONSTANT CURRENT.

1. **Electrical Stimulation.   Single Induction Shocks.—Apparatus.** —Frog, Daniell's cell, induction machine, two du Bois keys, five wires, flexible electrodes.

(*a.*) Arrange a cell and induction machine, for **single induction shocks** according to the scheme, Fig. 54.   Flexible

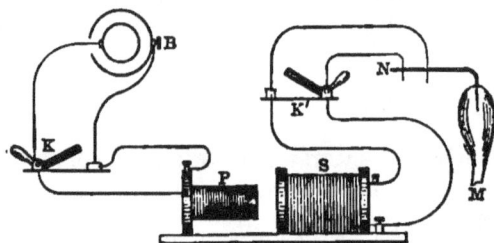

Fig. 54.—Scheme for Single Induction Shocks.—B, Battery; K, K', keys; P, primary, and S, secondary coil of the induction machine; N, nerve; M, muscle.

electrodes are fixed to the short-circuiting key (K') in the secondary circuit, and over them the nerve is to be placed.

(*b.*) Expose the sciatic nerve in a pithed frog, place the electrodes—preferably a pair fixed in ebonite, and so shielded that only one surface of their platinum terminals is exposed under it. Pull the secondary coil (S) far away from the primary (P), raise the short-circuiting key (K'), make and break the primary circuit by means of the key (K). At first there may be no contraction, but on approximating the secondary to the primary coil a single muscular contraction will be obtained, first with the break shock, and on approaching the secondary nearer to the primary coil, also with the make. The one is called a make and the other a break contraction. Record the results obtained.

2. Interrupted Current.

(a.) Arrange the induction machine so as to cause Neef's hammer to vibrate as directed in Lesson XXV, 2. On applying the electrodes to the sciatic nerve or gastrocnemius muscle, at once the muscle is thrown into a state of rigid spasm or continuous contraction, called tetanus, this condition lasting as long as the nerve or muscle is stimulated, or until exhaustion occurs.

3. Constant Current—Apparatus.—One, two, or three Daniell's cells, du Bois key, four wires and pair of electrodes, forceps, and nerve-muscle preparation.

(a.) Use two Daniell's cells. If two or more Daniell's cells be used, always connect them in series—i.e., the positive pole of one cell with the negative pole of the next one. Connect two wires, as in Fig. 55, to the free + and − poles of the battery (B), and introduce a du Bois key (K'), so as to short-circuit the battery circuit. Fix two shielded electrodes in the other binding-screws of the du Bois key, and having prepared a fresh nerve-muscle preparation, lay the divided sciatic nerve (N) across them, as shown in Fig 55.

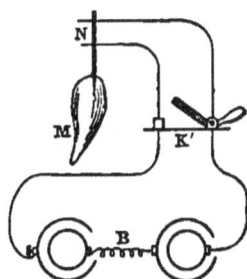

Fig. 55.—Scheme of constant current.—B, battery; K', short-circuiting key; N, nerve; M, muscle.

(b.) Make and break the current, and a single muscular contraction or twitch is obtained either at making or breaking, or both at making and breaking. Notice that if the key be raised to allow the current to flow continuously through the nerve, no contraction occurs, provided there be no variation in the intensity of the current. The electrodes may also be applied to the muscle directly.

(c.) Rapidly make and break the current by opening and closing the key, a more or less perfect *tetanus* is produced.

4. Dead Muscle and Nerve.—Immerse a nerve-preparation for a few minutes in water at 40° C. Both are killed, and none of the above stimuli cause contraction.

5. The Sartorius.—The student gets a clear idea of the shortening and thickening which occur when a muscle contracts by working with the sartorius muscle, because its fibres are arranged in a parallel manner.

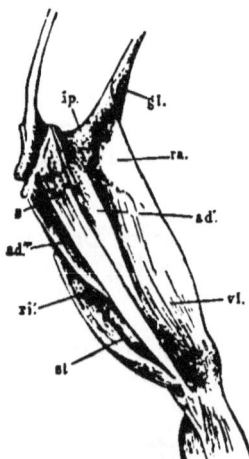

Fig. 56.—Muscles of the left leg of a frog seen from the front.—*ip*, ileo-psoas; *s*, sartorius; *ad'*, adductor longus; *vi*, vastus internus (see Figs. 50 and 51).

(a.) Pith a frog, skin it, lay it on its back, and dissect off the long narrow sartorius from the inner side of the thigh. Stretch it on a slip of glass (Fig. 56, *s*).

(b.) Stimulate the muscle first at its ends and afterwards at its centre or equator, as in Lesson XXVIII., 1,2, with (i.) a single induction shock, and (ii.) afterwards with an interrupted current. Observe the shortening and thickening, which are much greater in (ii.) than (i.) The muscle may be extended again, and stimulated as frequently as desired if it be kept moist.

6. Unipolar Stimulation—Apparatus.— Daniell's cell, induction machine, du Bois key, muscle-chamber, four wires.

(a.) Connect the Daniell to the primary coil of the induction machine either for single shocks or tetanus, introducing a du Bois key in the circuit. Connect one wire with the secondary coil, and attach it to one of the binding screws on the platform of the muscle-chamber, to which the nerve electrodes are attached. See that the battery and induction machine are perfectly insulated by supporting them on blocks of paraffin.

(*b.*) Prepare a nerve-muscle preparation, and arrange it in the muscle-chamber in the usual way, laying the nerve over the electrodes. One of the electrodes will therefore be connected with the secondary circuit.

(*c.*) Make and break the primary circuit, there is no contraction.

(*d.*) Destroy the insulation of the preparation by touching the muscle, or what does better, allow the brass support of the muscle to touch a piece of moist blotting-paper on the inner surface of the glass shade of the chamber. Every time the brass-binding of the shade is touched, or the brass support itself, the muscle contracts. Touch the secondary coil and contraction results.

LESSON XXIX.

RHEONOM—TELEPHONE EXPERIMENT— DIRECT AND INDIRECT STIMULATION OF MUSCLE—RUPTURING STRAIN OF TENDON—MUSCLE SOUND —DYNAMOMETERS.

1. **Fleischl's Rheonom.**—This instrument (Fig. 57) is very useful for showing du Bois-Reymond's law, that it is variations in the density of a galvanic current which excite a motor nerve. It consists of a square ebonite base, with a grooved circular channel in it, and two binding screws, with zinc attached, and bent over so as to dip into the groove, which is filled with a saturated solution of zinc sulphate. A vertical arm, with binding screws attached to two bent strips of zinc, moves on a vertical support.

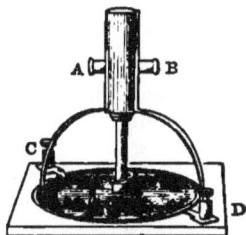

Fig. 57.—Fleischl's Rheonom.

(*a.*) Connect two Daniell's cells with the binding screws, A and B, introducing a du Bois key in one wire. Attach the electrodes, introducing a du Bois key to short-circuit them, to the binding screws, C and D. Fill the groove with a saturated solution of zinc sulphate.

(*b.*) Arrange the nerve of a nerve-muscle preparation in the usual way over the electrodes (Lesson XXVIII., 4). Pass a constant current through the nerve, observing the usual effects, viz., contraction at make or break, or both, but none when the current is passing. Then suddenly rotate the handle with its two zinc arms; this is equivalent to a sudden variation of the intensity of the current; the current, of course, continuing to pass all the time. The muscle suddenly contracts.

2. Telephone Experiment.

(*a.*) Arrange a nerve-muscle preparation with its nerve over a pair of electrodes. Connect the latter with a short-circuiting du Bois key. To the key attach the two wires from a telephone.

(*b.*) Open the short-circuiting key ; shout into the telephone, and observe that on doing so the muscle contracts vigorously.

3. Direct and Indirect Stimulation of Muscle.—When the stimulus is applied directly to the **muscle** itself, we have **direct stimulation** ; but when it is applied to the **nerve**, and the muscle contracts, this is **indirect stimulation** of the muscle.

(*a.*) Arrange a nerve-muscle preparation, and an induction machine for single or interrupted shocks (Lesson XXVIII., 1).

(*b.*) Test first the strength of current—as measured by the distance between the secondary and primary coils—which causes the muscle to contract when the stimulus is applied to the nerve—*i.e.*, for *indirect-stimulation.*

(*c.*) Then with the secondary still at the same distance from the primary, try if a contraction is obtained on stimulating the muscle *directly.* It will not contract, but make the current stronger, and it will do so.

4. Rupturing Strain of Muscle and Tendon.

(*a*.) Dissect out the femur and gastrocnemius with the tendo achillis of a frog. Fix the femur in a strong clamp on a stand, preferably one with a heavy base. To the tendo achillis tie a stout thread, and hang a scale pan on to it.

(*b*.) Into the scale pan place weights, and observe the weight required to rupture the tendon or muscle. Usually the muscle is broken first, and the weight added to the scale pan will be a kilo, more or less, according to the size of the frog.

(*c*.) Compare the rupturing strain of a frog's gastrocnemius which has been dead for twenty-four hours. A much less weight is required.

5. Muscle Sound.

(*a*.) Insert the tips of the index fingers into the auditory meatuses, forcibly contract the biceps muscles. A low rumbling sound is heard.

(*b*.) When all is still at night, firmly close the jaws, and especially if the ears be stopped, the sound is heard.

6. Dynamometers.

(*a*.) **Hand.**—Test the force exerted first by the right hand and then by the left, by means of Salter's dynamometer.

(*b*.) **Arm.**—Using one of Salter's dynamometers, test the strength of the arm when exerted in pulling, as an archer does when drawing a bow.

LESSON XXX.

INDEPENDENT MUSCULAR EXCITABILITY —ACTION OF CURARE—ROSENTHAL'S MODIFICATION—POHL'S COMMUTATOR.

1. Independent Muscular Excitability and the Action of Curare. —Curare paralyses the intra-muscular terminations of the motor nerves.—**Apparatus.**—Daniell's cell, induction machine, two keys, five wires, shielded electrodes, scissors, fine-pointed forceps, fine aneurism needle, or fine sewing needle fixed in a handle, with the eye free to serve as an aneurism needle, fine threads, pithing needle, 1 per cent. watery solution of curara in a glass-stoppered bottle, fine hypodermic syringe or glass pipette, frog.

(*a.*) Arrange the battery and induction machine for an interrupted current with a key in the primary circuit, and a du Bois key to short-circuit the secondary as in Lesson XXVIII., 2).

(*b.*) Pith a frog, destroying only its brain, and inject into the ventral or dorsal lymph sac one or two drops of a 1 per cent. watery solution of curara. The poison is rapidly absorbed. At first the frog draws up its legs, in a few minutes it ceases to do so, and will lie in any position in which it is put, while the legs are not drawn up on being pinched, and the animal lies flaccid and apparently paralysed.

(*c.*) Place the frog on its back. Expose the heart, and observe that it is still beating. Take care to lose no blood.

(*d.*) Expose the sciatic nerve on one or both sides.

(i.) Apply the shielded electrodes under them, and stimulate the *nerves* with tetanising shocks. There is *no contraction*.

(ii.) Apply the electrodes to the *muscles*, they *contract*.

Therefore, curara has paralysed the voluntary motor nerves, but not the muscles.

2. On what part of the Nerve does the Curare act?

(*a.*) Keep the induction apparatus as in the previous experiment.

(*b.*) Pith a frog, destroying only its brain. Carefully expose the sciatic nerve and the accompanying artery and vein on one side, *e.g.*, the *left*, taking great care not to injure the blood-vessels, which are to be carefully isolated for a short distance with a finder. Thread a fine aneurism needle with a fine silk thread. Moisten the thread with salt solution, and gently pass it under the sciatic artery. Withdraw the needle and ligature the artery. Instead of ligaturing merely the artery, it is better to isolate the sciatic nerve, and then to tie a stout ligature round all the other structures of the thigh. In this way none of the poison can pass by a collateral circulation into the parts below the ligature.

(c.) Inject a few drops of a 1 per cent. solution of curara into the ventral lymph sac, either by means of a hypodermic syringe or a fine pipette. In a short time the poison will be carried to every part of the body except the left leg below the ligature. Observe that the animal is rapidly paralysed, but if the non-poisoned leg (left) is pinched, it is drawn up, while the poisoned leg (right) is not.

(d.) Wait until the animal is thoroughly under the influence of the poison, and then expose *both sciatic nerves* as far up as the vertebral column and as far down as the knee.

(i.) Stimulate the *right* sciatic nerve. There is no contraction. Therefore the poison has acted either on nerve or muscle.

(ii.) Stimulate the *right* gastrocnemius muscle, it contracts. Therefore the poison has acted on some part of the nervous path, but not on the muscle.

(iii.) Stimulate the *left* sciatic *above* the *ligature*, the left leg contracts.

Observe that the part of the nerve above the ligature was supplied with poisoned blood, and has been under the influence of the poison, so that the nerve-trunk itself is not paralysed, as may be proved by stimulating any part of the *left* sciatic as far down as its entrance into the gastrocnemius. Stimulating any part of the left nerve causes contraction. Therefore, neither nerve-trunk nor muscle is affected. The nerve impulse is blocked somewhere, in all probability by paralysis of the terminations of the motor nerves within the muscle.

(e.) Apply several drops of a strong solution of curare to the *left* gastrocnemius, and after a time, stimulate the left sciatic nerve, there is no contraction, but on stimulating the muscle itself contraction takes place.

3. Rosenthal's Modification.

(*a*.) Prepare a frog as in the previous experiment, ligature the left leg—all except the sciatic nerve—and inject curare as before. After complete paralyses occurs, dissect out both legs with the nerves attached, but retain the legs, as in Fig. 58. Attach straw flags (N P and P) of different colours to the toes of both legs by pins, and fix both femora in muscle-forceps (F) with the gastrocnemii uppermost. Place the nerves (N) on the platinum points of du Bois-Reymond's electrodes (Fig. 47).

(*b*.) Arrange the induction apparatus as in Fig. 58. The primary coil is as before, but the terminals of the secondary coil are connected by two wires with the piers of a Pohl's commutator (Fig. 58) without cross-bars (H). Two other wires pass from two other binding screws of the commutator to the electrodes (N), while two *thin* wires pass from the other two binding screws (C) and their other ends are pushed through the gastrocnemii muscles. Place the commutator on a meat plate and fill its holes with mercury. The commutator enables the tetanising currents to be passed either through both nerves or both muscles. It is more convenient if the secondary circuit have a key, so that it may be short-circuited when desired.

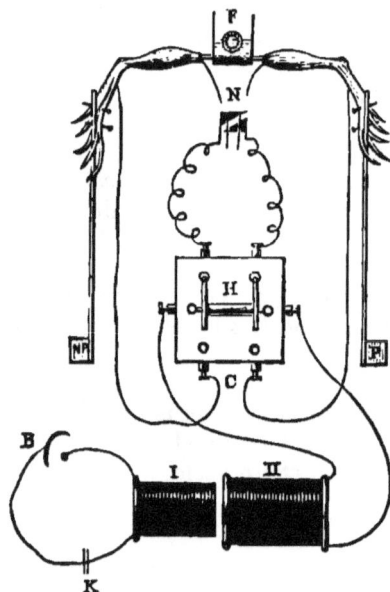

Fig. 58.—Scheme of the Curare Experiment.—B, Battery; I., primary, II., secondary spiral; N, nerves; F, clamp; N P, non-poisoned leg; P, poisoned leg; C, commutator; K, key.

(i.) Set Neef's hammer going, and turn the handle of the commutator so that the current passes through *both nerves;* only the non-poisoned leg (N P) contracts.

(ii.) Reverse the handle and pass the current through *both muscles, both contract.*

(iii.) Push the secondary spiral far away from the primary, and pass the current through *both muscles.* At first, if the spirals be sufficiently far apart, there is no contraction in either muscle. Gradually push up the secondary spiral, and notice on doing so that *the non-poisoned limb contracts* first, and that on continuing to push up the secondary spiral, both muscles ultimately contract (Rosenthal's Modification).

4. Pohl's Commutator (Fig. 59) is used for sending a current along two different pairs of wires, or for reversing the direction of the current in a pair of wires. It con-
sists of a round or square wooden or ebonite block with six cups, each in connection with a binding screw. Between two of these stretches a bridge insulated in the middle. The battery wires are always attached to the cups connected with this (1 and 2). When it is used to pass a current through different wires, the cross-bars are removed and wires are attached to all six cups, 3 and 4, 5 and 6.

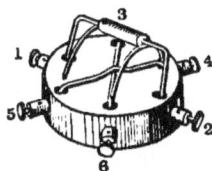

Fig. 59.—Pohl's Commutator with cross-bars in.

On turning the bridge to one side or other the current is sent through one or other pair of wires. To reverse the direction of a current, only one pair of wires, beside the battery wires, is attached to the mercury cups—*e.g.*, to 3 and 4, or 5 and 6, the cross-bars remaining in.

5. The student, if he desires, can prepare two nerve-muscle preparations, and dip the nerve of one (A) and the muscle of the other (B) into a solution of curara in two watch-glasses. On stimulating the nerve of A its muscle contracts; on stimulating the nerve of B its muscle does not contract, but the muscle contracts when it is stimulated directly. In A, although the poison is applied directly to the nerve trunk, the nerve is not paralysed.

11

LESSON XXXI.

THE GRAPHIC METHOD—MOIST CHAMBER —SINGLE CONTRACTION—WORK DONE.

1. **Recording Apparatus.**—For this purpose a revolving cylinder covered with smoked glazed paper, or other moving surface is required. The velocity of the moving surface is usually determined by recording simultaneously the vibrations of a tuning-fork, of known rate of vibration, or an electromagnetic time-marker. It does not matter particularly what form of recording drum is used, provided it moves smoothly and evenly, and is capable of being made to move at different rates as required. In Hawksley's form this is accomplished by placing the drum on different axles, moving at different velocities. In Ludwig's form (Fig. 60), this is done by moving a small wheel, *n*, on a large brass disc, D. Where a number of men have to be taught at once, one must have recourse to an arrangement of shafting, moved, say by a water-motor or turbine, and from which several drums can be driven by cords. Or one may use a small gas engine as the motive power, and cords passing over pulleys to move the drums.

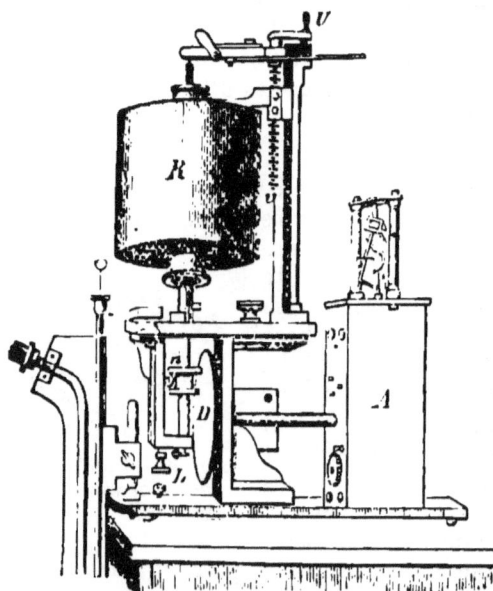

Fig. 60.—Ludwig's revolving cylinder, R, moved by the clock-work in the box, A, and regulated by a Foucault's regulator on the top of the box. The disc, D, moved by the clock-work, presses upon the wheel, *n*, which can be raised or lowered by the screw, L, thus altering the position of *n* on D, so as to cause the cylinder to rotate at different rates. The cylinder itself can be raised by the handle, *t*. On the left side of the figure is a mercurial manometer.

This is the arrangement adopted in the Physiological Department of Owens College, so that a large number of men can work at the same time, each being provided with recording apparatus for himself.

2. Cover the Cylinder with Paper.—The paper is glazed on one surface, and is cut to the necessary size to suit the drum. The drum can be removed from the clock-work or other motor which moves it, and is then covered with a strip of paper, the latter being laid on evenly to avoid folds. One edge of the paper is gummed, and slightly overlaps the other edge. Leave it for a few minutes until the gum dries. The paper has then to be **blackened**, by holding the drum and keeping it moving over a fan-tailed gas burner, or paraffin lamp—the former is preferable. Take care that the soot from the flame is deposited evenly and lightly, and see that it is not burned into the paper. The drum is then placed in position in connection with its motor.

3. General Rules to be observed with every Graphic or other Experiment.

(1.) In every experiment arrange the apparatus completely, cover the drum with paper, and smoke it, before beginning the dissection of the frog.

(2.) See that the secondary circuit is " short-circuited."

(3.) Test all the connections stage by stage as they are made.

(4.) Each tracing is to be inscribed with the name of the individual who made it, the date, and what it is intended to show, and any other particulars it is desired to record. It is then to be varnished, and the varnish allowed to dry.

4. Myographs.—Various forms are in use, but most of them consist of a lever which is raised by the contracting muscle, and so arranged as to record its movement on a smoked surface of paper or glass.

5. Moist Chamber (Fig. 61).—To prevent a nerve-muscle preparation from getting dry, it must be enclosed in a moist chamber, which is merely a glass shade placed over the preparation, while to keep the air and the preparation moist, the sides of the shade are covered with blotting-paper moistened with water.

6. Varnish for Tracings.—The tracing is simply drawn through the varnish and then hung up to dry.

(*a.*) A very good varnish consists of gum mastic dissolved to saturation in methylated spirit.

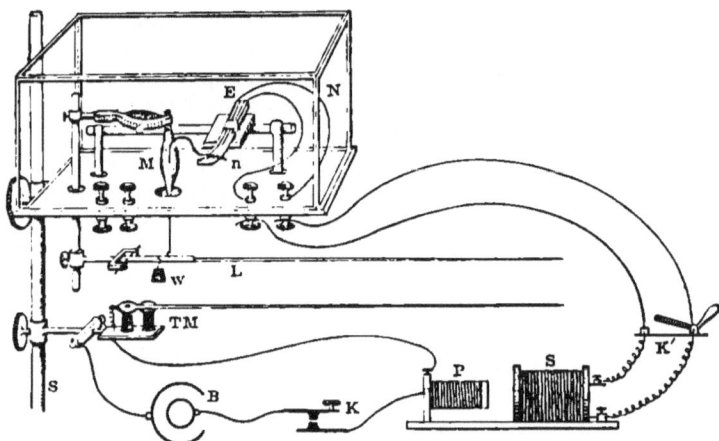

Fig. 61.—Moist Chamber.—N, Glass shade; E, electrodes; L, lever; W, weight; TM, time-marker; other letters as in previous figures.

(*b.*) Where a large quantity is used, and economy is an object, gum juniper may be used instead of mastic.

(*c.*) Dissolve 4 oz. of sandarac in 15 oz. of alcohol, and add half an oz. of chloroform.

7. Single Contraction or Twitch — Apparatus. — Recording drum, Daniell's cell, Morse key, induction machine, du Bois key, wires, electrodes, moist chamber and lever, moist blotting-paper, stout ligatures, hook, pins, lead-weight (20 grammes), frog, and the necessary instruments.

(*a.*) Arrange the recording apparatus and the drum to move slowly. Cover the drum with glazed paper, and afterwards smoke it over a gas flame, and fix it in position.

(*b.*) Arrange the apparatus as follows:—One Daniell's cell and a Morse key in the primary circuit, the secondary circuit of the induction machine short-circuited, and with wires to go to the binding screws on the platform of the moist chamber on the myograph (Fig. 62). [The muscle may be caused to contract either by stimulating it directly—in which case the electrodes are made of thin wires and merely pushed through the two ends of the gastrocnemius —or indirectly through the nerve. It is convenient to use the latter method (Lesson XXIX., 3).]

(c.) Make a nerve-muscle preparation, leaving the lower end of the femur in connection with the gastrocnemius, and cut away the tibia and fibula. With the point of a sharp pair of small scissors make a small hole in the tendo Achillis, and insert in it a hook like the letter S, made by bending a pin. Manipulate the nerve with a camel's-hair pencil. Arrange the preparation in the moist chamber by fixing the femur in the muscle clamp, and by means of a stout ligature thread attach the hook in the tendo Achillis to the writing-lever under the ebonite or wooden stage of the moist chamber. See that the muscle or ligature goes clear through the hole in the stage, and that the hook does not catch on anything. Adjust the height of the muscle clamp, so that the writing-lever is horizontal. Place the nerve over the electrodes, and cover the whole preparation with the glass-shade lined on three sides with moist blotting-paper. Load the lever near where the muscle is attached to it by a weight of 20 grammes or thereby, and make the lever itself horizontal. Arrange the point of the lever so that it writes on the cylinder. The writing-style on the tip of the lever may be made of very thin copper foil or parchment paper, fastened on to the lever with sealing-wax or telegraph composition.

8. According as the recording surface is stationary, or moving when the muscle contracts and raises the lever, either a vertical line or a curve will be made upon the paper. In the latter case the form of the curve will vary with the velocity of the drum. Arrange experiments for both.

A. Begin with the recording cylinder stationary.

(a.) Push the secondary coil far away from the primary, open the key in the secondary circuit, and make and break the primary circuit. There may be no contraction. Close the secondary circuit key.

(*b.*) Gradually approximate the secondary coil, open the short-circuiting key, and break the primary circuit by means of the Morse key in it. Observe when the first feeble contraction is obtained = **minimal contraction.** Make the primary circuit, there is no contraction. The break shock is, therefore, stronger than the make. Record under each contraction whether it is a make (M) or break (B) shock, and the distance in centimetres of the secondary from the primary coil. Move the drum a short distance with the hand, the lever will inscribe a horizontal or base line, the abscissa.

(*c.*) Gradually approximate the secondary spiral, and from time to time test the effect of the make and break shocks, after each test moving the cylinder with the hand, and recording the result as to M or B, and the distance in centimeters of the secondary from the primary coil. After a time a make contraction appears, and on pushing up the secondary coil the make contraction becomes as high as the break.

(*d.*) Increase the stimulus by bringing the secondary nearer the primary coil, and notice that the contractions do not become higher = **maximal contraction.** In each case keep the make and break contractions, obtained with each strength of current, close together. Their relative heights can then be readily compared.

B. Arrange the experiment as in **A**, but allow the **cylinder to revolve** at a moderate speed, 25-30 centimetres per second; the writing-style records an even horizontal line = the abscissa.

(*a.*) Select a strength of stimulus which is known to cause a contraction, and while the cylinder is revolving, cause the muscle to contract either by a make or break shock. Study the characters of the " **muscle-curve.**"

(*b.*) Vary the velocity of movement of the cylinder, and observe how the form of the curve varies with the variation in velocity of the cylinder.

(*c.*) Remove the tracings in **A** and **B**, and varnish them. In this case the moment of stimulation is not recorded.

9. The work done.—After the tracing of **B** is dry, from the abscissa draw vertical lines, or ordinates, and measure their height in millimetres. Measure the length of the lever, and from this calculate the actual amount of shortening of the muscle itself. Multiply this by the weight lifted, and the product is the work done expressed in gram-millimetres.

10. The Crank Myograph (Fig. 62) is fixed on a suitable support, so that it can be adjusted to any height desired. The experiment is arranged in exactly the same way as for **7.**

(*a.*) Use one hind-limb of a pithed frog; pin the femur firmly to the cork plate of the myograph covered with blotting-paper moistened by normal saline, the tibia being in line with the writing-lever. Or take the pithed frog, lay it on the frog-plate of the myograph, expose the gastrocnemius, and proceed as above. Detach the tendo Achillis, tie a stout ligature to its sesamoid bone, and fix the liga-

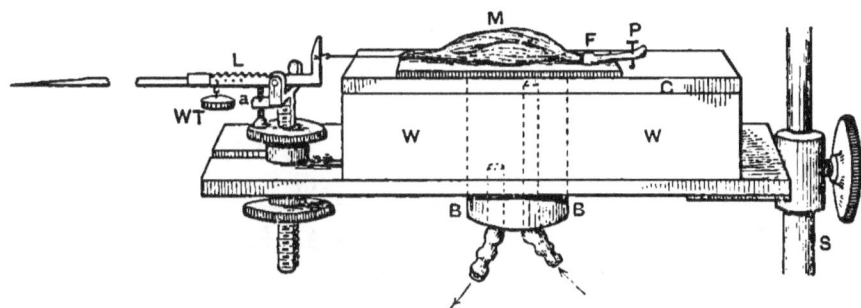

Fig. 62.—Crank Myograph.—W, W, Block of wood; M, muscle; F, femur; P, pin to fix F; L, lever; WT, weight; a, screw for after-load; C, cork; B,B, brass box. (In this figure the fulcrum should be at the angle of the crank.)

ture to the short arm of the lever, add a weight of 20 grammes to the lever, and see that the lever itself is horizontal. Thrust two fine wires—the electrodes—from the du Bois key in the secondary coil, through the upper and lower end of the gastrocnemius muscle.

(*b.*) Arrange the style of the lever so that it writes on the cylinder, and repeat the experiments of either **A** or **B**, or both.

(c.) Use different weights—5—20—50 grammes—and observe how the form of the curve varies on increasing the weight attached to the lever.

11. After-load.—In the crank-myograph, under the lever is a screw on which the horizontal arm of the bell-crank rests (Fig. 62, a), so that the muscle is loaded only during its contraction.

12. Interrupted Current.—If instead of single shocks, the induction apparatus be so arranged that Neef's hammer is in action, then on stimulating the muscle or nerve, **tetanus** is obtained (Lesson XXVIII, 3), and the curve of a tetanised muscle recorded.

LESSON XXXII.

ANALYSIS OF A MUSCULAR CONTRACTION —PENDULUM MYOGRAPH—SPRING MYOGRAPH—TIME-MARKER— DEPREZ' SIGNAL.

1. Muscle Curve by the Pendulum Myograph.

(a.) Cover the oblong glass plate with glazed paper, smoke its surface, and fix it to the pendulum. The plate must be so adjusted that the pendulum on being set free from the "detent" (Fig. 63, C) shall be held by the "catch" (C). Test this.

(b.) Arrange the induction apparatus for single shocks as in Fig. 63, but short-circuit the secondary circuit, interposing in the primary circuit the trigger-key or knock-over key of the pendulum myograph (Fig. 63, K').

(c.) Make a nerve-muscle preparation, fix the femur in the clamp, attach the tendo Achillis to the writing-lever (S), and place the nerve over the electrodes in the ordinary muscle moist chamber. Adjust the moist chamber on its stand and raise it to a suitable height. Load the lever with 20 grammes, and direct its point to the side to which the pendulum swings. Fix the pendulum with the detent, and adjust the writing-style of the lever on the smoked surface. Connect the secondary spiral either with the

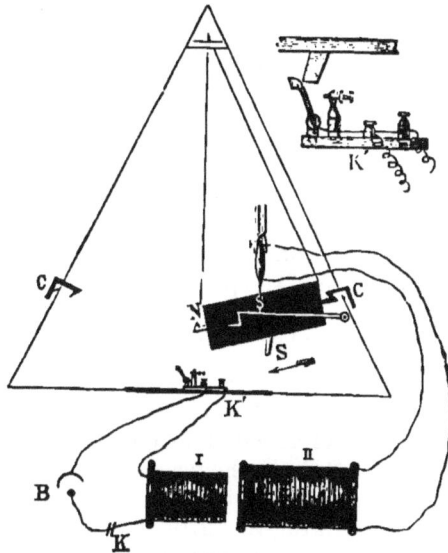

Fig. 63.—Scheme of the arrangement of the pendulum.—B, Battery; I., primary, II., secondary spiral of the induction machine; S', tooth; K', key; C, C, catches; K' in the corner, scheme of K'; K, key in primary circuit.

muscle directly, or preferably, with the electrodes on which its nerve rests, introducing a short-circuiting key. In Fig. 63 this is omitted, and the wires of the secondary circuit go direct to the muscle. After un-short-circuiting, with the hand break the primary circuit to make certain that a contraction occurs on breaking the primary circuit.

(d.) See that the trigger-key (K') of the pendulum (in the primary circuit) is closed, and the key in the secondary circuit open. Allow the pendulum to swing; as it does so, it knocks open the key in the primary circuit and breaks the current, thus inducing a shock in the secondary circuit, whereby the muscle is stimulated and caused to record its contraction or muscle-curve on the smoked surface.

(e.) Take the abscissa—i.e., the base line. Rotate the stand on which the moist chamber is supported, so as to withdraw the writing point of the lever from the recording surface. Bring the pendulum back to the detent, adjust the writing-style, close the trigger-key, and keep the secondary circuit short-circuited by closing the du Bois key. Allow the pendulum to swing. This records the abscissa or base line.

(*f.*) Ascertain the **latent period** as follows :—Bring the pendulum to the detent, short-circuit the secondary current, and withdraw the writing-style as in (*e.*) Close the trigger-key of the pendulum, and with a finger of the left hand keep it closed. Allow the lever to touch the glass plate in its original position, and with the right hand bring the knife-edge of the pendulum in contact with the trigger-key so as just to open it. Thus a vertical line is inscribed on the stationary plate, which indicates the moment of stimulation.

(*g.*) Remove the muscle lever, place the pendulum in the detent, close the trigger-key, take a tuning-fork vibrating, say, 120 or 250 double vibrations per second, and adjust its writing-style in the position formerly occupied by the style of the muscle lever. Set the fork vibrating, either electrically or by striking it. Allow the pendulum to swing, when the vibrating tuning-fork will record the **time curve** under the muscle curve (Fig. 64, 250 DV). All the conditions must be exactly the same as when the muscle curve was taken.

(*h.*) Remove the paper, varnish the tracing, hang it up to dry, and next day measure the tracing. Bring ordinates vertical, *a', b', c',* to the abscissa, and measure the "**latent period**" (Fig. 64, A), the duration of the shortening (B), the phase of relaxation (C), and the contraction remainder.

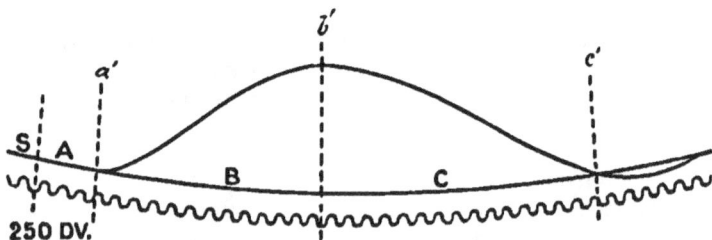

Fig. 64.—Pendulum Myograph Curve.—S, Point of stimulation; A, latent period; B, period of shortening, and C, of relaxation.

2. **Spring Myograph** (Fig. 65).—The arrangements are exactly the same as for the pendulum myograph, the trigger-key of the myograph being placed in the primary circuit. The instrument must be raised on blocks.

(*a.*) Cover the slip of glass with glazed paper, smoke it, and fix it in the frame. Push the plate to one side by means of the rod attached to it, and fix it by means of the catch.

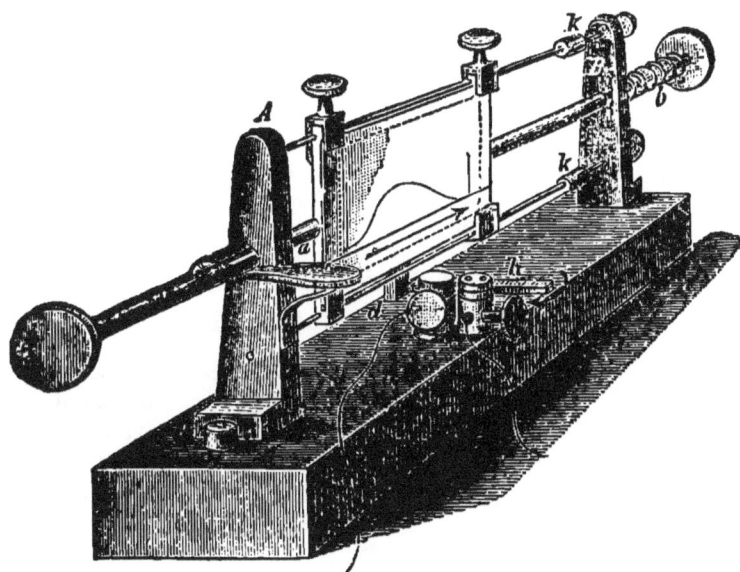

Fig. 65.—Spring Myograph.

Close the trigger-key (*h*), and introduce it into the primary circuit of the induction machine.

(*b.*) Make a nerve-muscle preparation, and arrange it to write on the glass plates as directed in 1 (*c.*) Remember to un-short-circuit the secondary circuit.

(*c.*) Press on the thumb-plate (*a*), thus liberating the spring, when the glass plate moves swiftly to the other side, and in doing so the tooth (*d*) on its under surface breaks the primary circuit, and the muscle-curve is recorded.

(*d.*) Short-circuit the secondary circuit, push back the glass plate, and fix it with the catch ; close the trigger-key, and shoot the glass plate again. This records the abscissa, or horizontal line.

(*e.*) Remove the moist chamber, and take the time curve. Push the glass plate back again, and secure it by the catch ; close the trigger-key—in order that the conditions may be exactly the same as before—strike a tuning-fork, vibrating, say, 120 double vibrations per second, and when it is vibrating adjust its writing-style under the abscissa. Shoot the glass plate again, and the time curve will be recorded.

(*f.*) Remove the tracing, fix it and measure it out, determining the length of the latent period and the duration of the contraction, and of its several parts.

3. Study the **improved form** of this instrument recently introduced by du Bois, in which the glass plate is set free, and the tuning-fork vibrations are recorded simultaneously when a handle is pressed. It is a most elegant piece of apparatus, and has a beautiful mechanism for adjusting the writing-styles for the muscle and abscissa.

4. On a Revolving Cylinder.

(*a.*) Arrange the drum to move at the fastest speed.

(*b.*) Arrange an induction machine for single shocks, the secondary circuit to be short-circuited, and arranged to stimulate a nerve attached to a muscle placed in a moist chamber, as directed for the foregoing experiments. Into the primary circuit introduce besides the Morse key, an electro-magnet with a marking lever (Figs. 61, 66, *e*), and cause its point to write exactly under the muscle lever. Arrange, with its point exactly under the other two, a Deprèz' chronograph or signal, in circuit with a tuning-fork of known rate of vibration, and driven by means of a battery (Fig. 68).

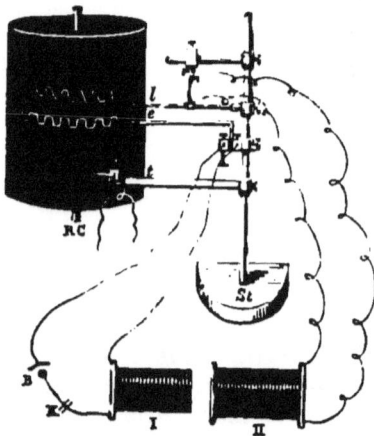

Fig. 66.—Arrangement for Estimating the Time-Relations of a Single Muscular Contraction.—B, battery; K, key in primary circuit; I., primary, II., secondary spiral, without a short-circuiting key; *l*, muscle lever; *e*, electro-magnet in primary circuit; *t*, electric signal; *St*, support; RC, revolving cylinder.

The three recording levers are all fixed on the same stand, which should preferably be a tangent one—*i.e.*, the rod bearing the recording styles can by means of a handle be made to rotate so as to bring the writing-styles in contact with the recording surface. This avoids the overlapping of the time curve which otherwise happens.

On un-short-circuiting the secondary circuit, and breaking the primary one, the muscle contracts, and at the same time the style of the electro-magnet is attracted and records the exact moment of stimulation (Fig. 61).

5. Time-Recorder (Fig. 67).—This is merely an electro-magnet introduced into an electric circuit, and the magnet, B, is so arranged as to attract a writing-style, L.

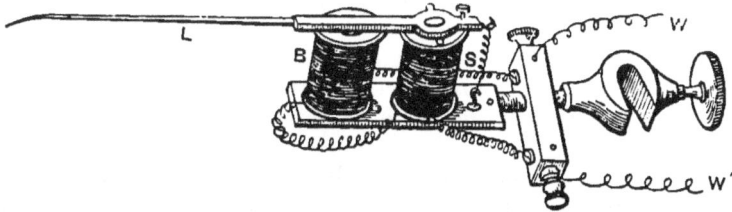

Fig. 67.—Time-Marker.—L, Lever ; B, electro-magnet bobbins; S, support; W, W', wires.

6. Deprèz' Signal (Fig. 68).—This small electro-magnet has so little inertia that, if it be introduced into an electric circuit, its armature, which is provided with a very light writing point, vibrates simultaneously with the vibrations of an electric tuning-fork introduced into the same circuit. Arrange the signal

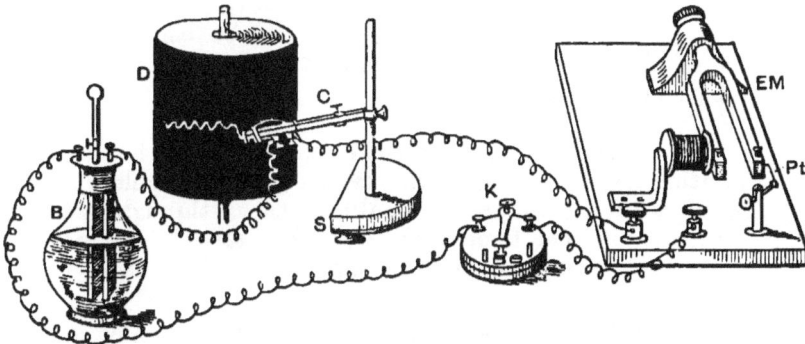

Fig. 68.—Signal and Vibrating Tuning-Fork in an Electric Circuit.—D, Drum ; C, signal ; EM, electric tuning-fork ; Pt, platinum contact.

and tuning-fork as in Fig. 68. The drum must move more rapidly, the more rapid the vibrations of the tuning-fork used.

LESSON XXXIII.

INFLUENCE OF TEMPERATURE, LOAD, VERATRIA ON MUSCULAR CONTRACTION.

1. Influence of Temperature on Muscular Contraction.

(*a.*) Arrange the experiment with a crank-myograph as in Lesson XXXI., 10, but do not remove the skin of the leg. Take a tracing at the normal temperature.

(*b.*) Alter the height of the drum or that of the myograph. Place ice upon the skin over the gastrocnemius for some time, and then take another tracing, noting the differences in the result, the contraction being much longer.

(*c.*) Adjust a piece of wire gauze over the leg, and allow it to project beyond the end of the plate of the myograph. Heat the gauze with a spirit-lamp. Take a tracing. The contraction is shorter than in 1 (*b.*) Do not overheat the muscle.

(*d.*) A piece of thin gas piping can be bent, and the muscle laid on it. Water of various temperatures can then be passed through it.

(*e.*) The muscle may be attached to an ordinary horizontal writing-lever. Surround the muscle with a double-walled box, with an inflow and outflow tube, through which water at different temperatures can be passed. A delicate thermometer is placed in the chamber with the muscle.

(*f.*) Perhaps for the purpose of the student the most convenient method is to allow the muscle to rest on a small circular brass box, fitted into the wooden plate of the crank-myograph. The box (B, B) is provided with an inflow and an outflow tube, through which water of the desired temperature can be passed (Fig. 62).

2. Influence of the Load on the Form of the Curve.

(*a.*) Arrange an experiment with the pendulum myograph as in Lesson XXXII., 1, using either a muscle-lever or a crank-myograph.

(*b.*) Take a tracing with the muscle weighted with the lever only.

(*c.*) Then load the lever successively with different weights

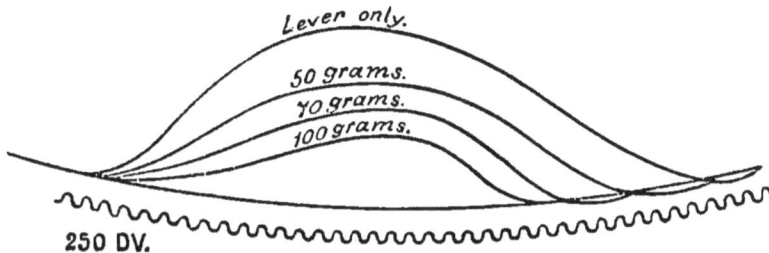

Fig. 69.—Pendulum Myograph Curves showing the Influence of the Load on the Form of the Curve.

(5 . . . 20 . . . 50 . . . 70 . . . 100 grammes), and in each case record a curve and observe how the form of the curve varies (Fig. 69).

(*d.*) In each case record the abscissa and time curve with the usual precautions.

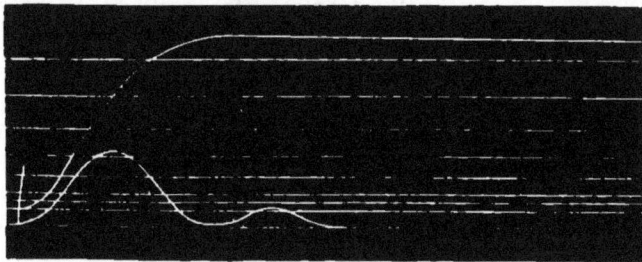

Fig. 70.—Veratria Curve (Upper).   Normal Muscle Curve (Lower).

3. Influence of Veratria on Contraction.

(*a.*) Destroy the brain of a frog, and inject into the ventral lymph sac 10 minims of a freshly-prepared 0·1 per cent. solution of veratria.

(*b.*) Arrange the induction machine for single shocks.

(*c.*) Make a nerve-muscle preparation, fix it in a moist chamber, and arrange the muscle lever to record its movements on a slowly revolving drum. Take a tracing, observing the long drawn-out form of the curve, and how long the muscle takes to relax.

(*d.*) The direct action of veratria on muscular tissue may also be studied by the apparatus described in Lesson XXXVII., and by this method it is easy to compare the form of the curve before and after the action of the poison (Fig. 70).

LESSON XXXIV.

ELASTICITY AND EXTENSIBILITY OF MUSCLE.

1. **Elasticity of Muscle.**

(*a.*) Dissect out the gastrocnemius of a frog with the femur attached, clamp the femur, attach the tendon to the light writing muscle lever of the moist chamber, and fix a scale pan to the lever. Neglect the weight of the pan, and see that the lever writes horizontally on a stationary drum.

(*b.*) Place in the scale pan successively different weights (10, 20, 30, 40 . . . 100 grammes). On placing in 10 grammes the lever will descend, remove the weight and the lever will ascend. Move the drum a certain distance, and add 20 grammes to the scale pan. This time the vertical line drawn is longer, indicating greater extension of the muscle by a greater weight, but nevertheless the muscle lever will rise to its original height on removing the weight. Repeat this with other weights. With the heavier weights care must be taken that everything is securely clamped. If the apices of all the lines obtained be joined, they form a *hyperbola*. The muscle therefore has not a large amount of elasticity, *i.e.*, it is easily extended by light weights, and on removal of the weight it regains its original length, so that its elasticity is said to be very perfect. The *hyperbolic curve* obtained shows further that the increase in length is not directly proportional to the weight; but it diminishes as the weights increase.

(c.) Repeat the same observation with a thin strip of *india-rubber*. In this case equal increments of weight give an equal elongation, so that a line joining the apices of the vertical lines drawn after each weight is a *straight line*.

2. The Extensibility of Muscle is Increased during Contraction, its Elasticity is Diminished.

(a.) Arrange a muscle in a moist chamber, connecting it to a lever to record on a drum, and adjust an interrupted current to stimulate the muscle, either directly or indirectly.

(b.) Load the lever with 50 grammes, and in doing so allow the drum to move slowly. Remove the load and observe the curve obtained.

(c.) Tetanise the muscle, and while it is contracted to its greatest extent, again load the lever with 50 grammes while the drum is in motion, and remove the load. Observe the curve.

(d.) Compare the two curves. The second curve will, of course, begin higher, but notice that its absolute descent is greater than the first curve, and that it does not rise to the horizontal again.

(e.) It is better to begin the experiment with the drum stationary, and then to record the tracing with the drum in motion.

(f.) A better curve is obtained by using a long counter-poised lever attached to the muscle, which writes on a very slow-moving drum. A weight is made to travel along the lever by means of two pulleys with an endless string.

LESSON XXXV.

FATIGUE OF MUSCLE.

1. Fatigue of Muscle.

(*a.*) Arrange an induction apparatus for single shocks, but introduce into the primary circuit in addition to the du Bois key a trigger-key, the latter fixed to a stand, and so placed that a tooth on the edge of the drum can knock it over, and thus break the primary current as required. Or attach to the edge of the under surface of the drum a short style; a strong pair of bull-dog forceps clamped on to it does perfectly well.

(*b*). Make a nerve-muscle preparation,

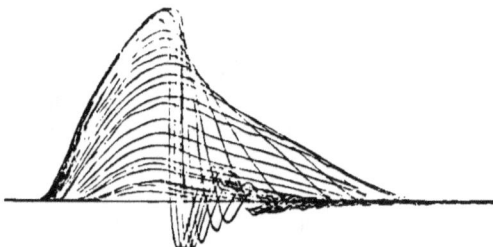

Fig. 71.—Fatigue Curve.—The sciatic nerve was stimulated with maximal induction shocks and every fifteenth contraction recorded.

clamp the femur, and adjust the preparation in the moist chamber, or a crank-myograph — the whole to be supported on a tangent stand. Attach the muscle to a writing-lever to record on a revolving drum.

(*c.*) Close the trigger-key, and on allowing the cylinder to revolve, the style knocks it over, breaks the primary circuit, and induces a shock in the secondary. Immediately short-circuit the secondary circuit, close the trigger-key and unshort-circuit the secondary circuit, and allow the drum to revolve. Repeat this until the muscle is fatigued. Record only every fifteenth contraction. In this way the muscle is always stimulated at the same moment, and the various curves are superposed, and can be readily compared (Fig. 71).

(*d.*) The best plan is to fix a platinum style on the spindle of the drum, and as it revolves it comes in contact with another piece of platinum introduced into the primary circuit, and fixed to the base of the drum-support, so that a break shock is obtained each time the drum revolves.

(*e.*) Observe how rapidly the height of the curves falls, while their duration is longer. In nearly every case fatigue curves from muscle show a "stair-case" character (Fig. 71), the second curve being higher than the first one, and the third than the second.

2. Instead of recording on a moving surface, a stationary one may be used (or a very slow-moving drum, 1 mm. per sec.), either a drum or a flat glass surface, while the muscle may be attached to a crank-myograph or to Pflüger's myograph (Fig. 72).

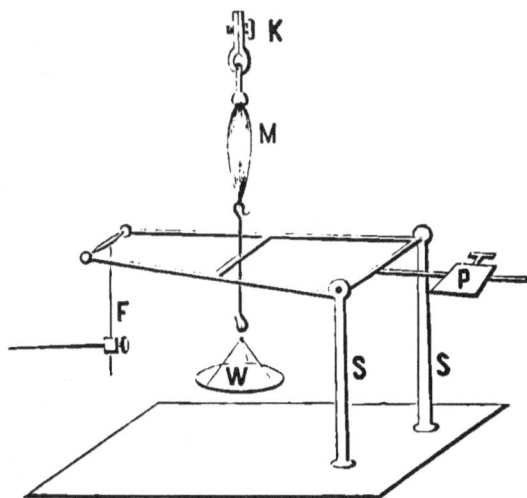

Fig. 72.—Pflüger's Myograph.—K, Clamp; M, muscle; W, weight in scale pan; F, writing-style; P, counterpoise of lever; S, S, supports.

(*a.*) Arrange the experiment as before, but adjust the muscle in a Pflüger's myograph, the primary circuit being still broken by a revolving drum.

(*b.*) When the muscle contracts, it merely makes a vertical scratch on the smoked glass plate of the myograph. Move the plate with the hand a little distance after each contraction. A series of vertical lines is obtained, and a straight line will join the apices of them all.

3. Seat of Exhaustion in Nerve and Muscle.

A.—(*a.*) Arrange an induction machine for interrupted shocks. Connect the secondary coil with a Pohl's commutator without cross-bars.

(*b.*) Arrange a Daniell's cell connected to N.P. electrodes, and short-circuited for a constant current—the "polarising current" (Lesson XXXVIII., 2).

(*c.*) Dissect out two nerve-muscle preparations (A and B) from a frog, clamp both femurs in one clamp, and attach straw-flags of different colours to both legs (Fig. 58). Lay both nerves over a pair of du Bois electrodes. Arrange a glass shade, lined in part with moist blotting-paper to cover them and keep them moist.

(*d.*) Attach the electrode wires to two of the binding screws of the commutator, turning the handle, so that the current can be passed through both nerves when desired.

(*e.*) Arrange the nerve of B on the N.P. electrodes placed between the tetanising current and the muscle, so that the – pole is next the muscle. Pass an interrupted current through *both nerves*, A will become tetanic, while B remains quiescent; the impulse cannot pass because of the effect of the "polarising current" producing electrotonus.

(*f.*) Continue to stimulate until the tetanus in A ceases. Break the polarising current, B becomes tetanic.

As both nerves have been equally stimulated, both are equally fatigued. As B becomes tetanic, the seat of the fatigue is not in the nerve trunk.

B.—(*a.*) Arrange an induction coil and commutator as before.

(*b.*) Prepare a nerve-muscle preparation, with a straw-flag as before, and place its nerve over the du Bois electrodes attached to the commutator. Pass two fine wires through the gastrocnemius, and attach them to the other two binding screws of the commutator.

(*c.*) Tetanise the nerve until the tetanus ceases. Then reverse the commutator and stimulate the muscle. It contracts. Therefore, in this case, the seat of fatigue is not in the muscle.

LESSON XXXVI.

PRODUCTION OF TETANUS—METRONOME —THICKENING OF A MUSCLE.

1. **Production of Tetanus.**—The tetanising current may be made and broken with the hand, by Neef's hammer, or by means of a vibrating rod. **Apparatus.**—Daniell's cell, five wires, flat spring, cup of mercury in a wooden stand, induction machine, du Bois key, moist chamber and muscle lever, recording drum moving at a medium speed.

(*a.*) Arrange the experiment as in Fig. 73; the induction machine for single shocks, short-circuiting the secondary circuit. Place in the primary circuit the flat metallic spring, held in a wooden clamp. One end of the spring has a needle fixed at right angles to it, which dips into a cup of mercury fixed in a wooden stand. The needle hangs just above the mercury cup when the spring is at rest, but dips in and out of the mercury when it vibrates. The clamped end of the spring is connected with the battery, while the mercury cup is connected with the induction machine. Cover the mercury with alcohol and water (1 : 3) to prevent oxidation, and to keep the resistance more uniform.

(b.) Make a nerve-muscle preparation, place it in a moist chamber, attach it to a writing-lever, weight the latter with 15 grammes, and cause it to write on a drum revolving at a moderate rate.

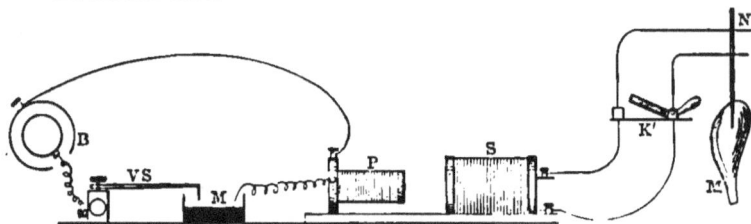

Fig. 73.—Scheme of Arrangement for Tetanus.—VS, Vibrating spring; M, cup for mercury. Other letters as before.

(c.) Fix the flat spring firmly in the clamp, with ten inches projecting. Allow the drum to revolve, set the spring vibrating, and while it is doing so, open the key in the secondary circuit, and before the spring ceases to vibrate short-circuit the secondary current.

(d.) Shorten the vibrating spring somewhat, and repeat the same experiment, making the tracing follow the previous one.

(e.) Make several more tracings, always shortening the spring, until the tracing no longer shows any undulations, until the spasms are completely fused—i.e., until it has passed from the phase of " incomplete " to " complete tetanus."

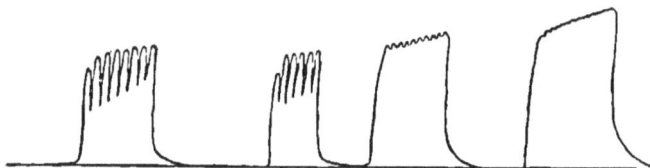

Fig. 74.—Curves obtained in previous experiment, the last almost a complete tetanus.

(f.) Finally take a tetanus-curve by introducing Neef's hammer instead of the vibrating flat spring.

(g.) Study the tracings. At first the tracings will be indented, but gradually there will be more and more fusion of the teeth until a curve unbroken by depressions is obtained. In the curve of complete tetanus the ascent is at first steep, then slightly more gradual, speedily reaching a maximum, when the lever practically records a horizontal line parallel to the abscissa. When the current is shut off the descent is very steep at first, and towards the end very slow.

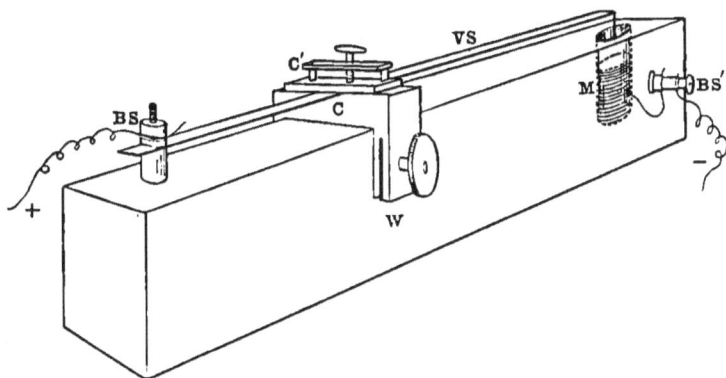

Fig. 75.—Interrupter for Tetanus.—W, Wooden block; VS, vibrating spring; BS and BS′, binding screws; C, movable clamp; C′, clamp to fix spring; M, cup of mercury.

2. Instead of using the spring held in a clamp, a convenient form is shown in Fig. 75. If desired the spring can be kept vibrating by an electro-magnet.

3. **Interruption by a Metronome.**—Instead of the vibrating rod or Neef's hammer, introduce into the primary circuit a metronome, provided with a wire which dips into a mercury cup introduced into the primary circuit. Vary the rate of vibration of the metronome, and observe the effect on the muscle-curve.

4. **Thickening of a Muscle during Contraction.**

(a.) Arrange a Marey's tambour to write on a pendulum myograph (Fig. 76).

(b.) Fix Marey's *pince myographique* so as to compress the adductor muscles between the thumb and the metacarpal bone of the index finger, keeping the two arms together with an elastic band. Or use a pair of toy bellows, to the arms of which plate-like electrodes are fitted and connected with binding screws. Keep the handles of the bellows pressed upon the adductor muscles by means of an elastic band. Connect the receiving tambour of the pince, or the nozzle of the bellows with the recording tambour, introducing a valve, or T-tube with a screw clamp, into the connecting elastic tube, to regulate the pressure of the air within the system of tubes.

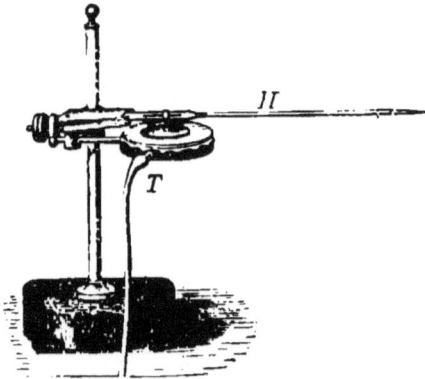

Fig. 76.—Marey's registering tambour. Metallic Capsule, T, with thin india-rubber stretched over it, and bearing an aluminium disc, which acts upon the writing-lever, H.

(*c.*) Arrange an induction machine with the trigger-key of the pendulum myograph in the primary circuit, and the pince or bellows in the secondary. Take a tracing in the ordinary way. The time relations of the contraction are determined in the manner already stated (Lesson XXXII).

LESSON XXXVII.

TWO SUCCESSIVE SHOCKS—ACTION OF DRUGS ON EXCISED MUSCLE.

1. **Two Successive Shocks.**—Use either the pendulum or spring myograph, preferably the former. Either instrument must be provided with two trigger-keys which are capable of being adjusted at any required distance from each other, and opened by the moving recording plate of the instrument.

(*a.*) Charge four Daniell cells, and connect two with one induction machine and two with another, introducing one trigger-key in the primary circuit of one induction machine, and the other trigger-key in the primary circuit of the other one. If the pendulum myograph be used, let the trigger-keys be about 5–8 centimetres apart. Connect the two secondary coils by a wire stretching between the two adjacent terminals. The other two terminals are connected with a short-circuiting du Bois key, from which the electrodes proceed.

(*b.*) Arrange a muscle or nerve-muscle preparation either in a moist chamber or on a crank myograph; place the electrodes under the nerve or stimulate the muscle directly.

(*c.*) Take a tracing in the ordinary way, after unshort-circuiting the secondary coils, and seeing that both primary circuits are closed. On discharging the instrument, first the one key and then the other is opened. It is necessary to ascertain beforehand that both break shocks are nearly of the same intensity. Take a series of tracings, gradually diminishing the distance between the two trigger-keys. In each case record the movement of stimulation with each trigger-key.

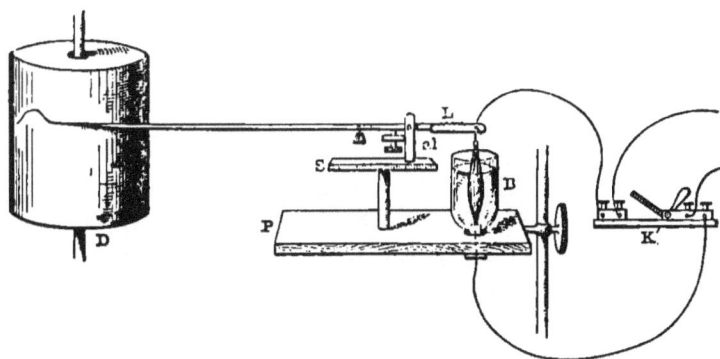

Fig. 77.—Wild's apparatus for studying the action of poisons on muscle.—
D, Drum; P, platform; S, stand; al, after-load; L, lever; B, bottle with muscle; K', key.

2. **Wild's Apparatus** consists of a glass cylinder made by inverting the neck end of a two-ounce phial. The neck is fitted

with a cork, the upper end is open (Fig. 77, B). A wire connected with a key (K') short-circuiting the secondary coil of an induction machine, perforates the cork. Arranged above is a light lever (L) provided with an after-load (al), and moving on an axis, the short arm projecting over the mouth of the jar. The whole arrangement is fixed to a platform (P), with an adjustable stand (S) bearing the fulcrum of the lever and the after-load. The cork must be renewed with each new drug used.

(a.) Dissect out the gastrocnemius, divide the femur with the gastrocnemius attached just above the attachment of the latter, and the tibia below the knee-joint. Pass a fine metallic hook through the knee-joint or its ligaments, and attach it to the projecting hook of fine wire fixed to the short arm of the lever. Fix the tendo achillis to a hook connected with the wire passing through the cork in the neck of the glass cylinder.

(b.) Fill the glass cylinder—which encloses the muscle— with normal saline. Stimulate the muscle directly with break shocks, and take a tracing.

(c.) Remove the normal saline with a pipette, and replace it with a solution of the drug whose action you wish to study—e.g., veratria 1 in 5000, or barium chloride 1 in 1000. Study the veratria tracing (Fig. 70).

--

LESSON XXXVIII.

DIFFERENTIAL ASTATIC GALVANOMETER —NON-POLARISABLE ELECTRODES— SHUNT—DEMARCATION AND ACTION CURRENTS IN MUSCLE.

ELECTRO-MOTIVE PHENOMENA OF MUSCLE AND NERVE.

1. **Thomson's High Resistance Differential Astatic Galvanometer.**

(*a.*) Place the galvanometer (Fig. 78) upon a stand where it is unaffected by vibrations—*e.g.*, on a slate slab fixed into the wall at a suitable height, or on a solid stone pillar fixed in the earth, taking care that no iron is near.

(*b.*) Place the galvanometer so that it faces *west—i.e.*, with the plane of the coils in the magnetic meridian, the magnetic meridian being ascertained by means of a magnetic needle. As the galvanometer is a differential one, to convert it into a single one, connect the two central binding screws on the ebonite base, by means of a copper wire.

(*c.*) By means of the three screws attached to the ebonite base, level the galvanometer.

(*d.*) Allow the mirror attached to the upper needle to swing freely. Take off the glass cover and steadily raise the small milled head on the top of the upper coils, which frees the mirror. Replace the glass shade.

(*e.*) Place the scale (Fig. 79) also in the magnetic meridian and 1 metre from the mirror,

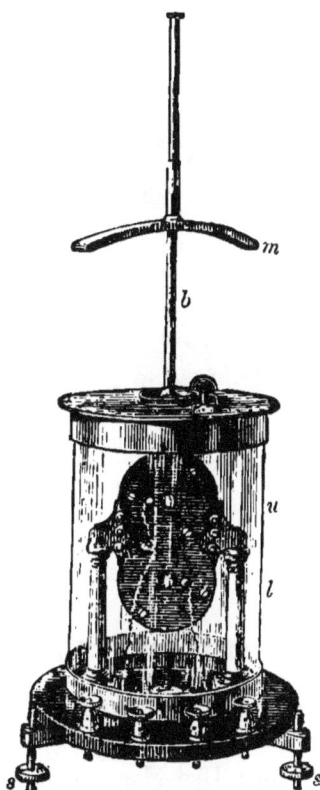

Fig. 78.—Sir William Thomson's reflecting galvanometer. —*u*, Upper, *l*, lower coil; *s*, *s*, levelling screws; *m*, magnet on a brass support, *b*.

taking care that it is at the proper height. An improved form of scale with several adjustments has recently been introduced. Instead of a mere slit in the scale, it is better to fix in it a thin wire, and by means of a lens of short focal distance to bring the image of the wire to a focus in the middle of the illuminated disc of light reflected from the mirror upon the scale.

(*f.*) Light the paraffin lamp, darken the room, and see

that the centre of the scale, its zero, the slit in the scale,
the flame of the lamp, and the centre of the mirror are all
in the same vertical plane, so that a good light is thrown
on the mirror in order to obtain a good image on the scale.
The lamp used is an ordinary
paraffin one provided with a
copper chimney with a plane
glass in front and a concave
mirror behind. When lighted,
the edge of the flame is turned
toward the slit.

(*g.*) Make the needle all but
astatic, by means of the magnet
attached to the bar above the
instrument. The needle is *most
sensitive when it swings slowly.*

Fig. 79.—Lamp and Scale for
Thomson's Galvanometer.

(*h.*) Test the sensitiveness of
the galvanometer by applying
the tips of two moist fingers to the two outer binding
screws of the instrument, when at once the beam of light
passes off the scale.

2. To Make Non-polarisable Electrodes.—One may use the
old form as invented by du Bois-Reymond, the simple tube
electrodes, or the " brush-electrodes " of v. Fleischl.

(*a.*) Use glass tubes about 4 cm. long, and 5 mm. in
diameter, tapering somewhat near one end, and see that they
are perfectly clean.

(*b.*) Plug the tapered end of the glass tube with a plug of
china clay, made by mixing kaolin into a paste with normal
saline. Push the clay into the lower third or thereby of the
tube to plug it, by means of a fresh cut piece of wood or
thin glass rod ; allow part of the clay to project beyond the
tapered end of the tube (Fig. 80, *t, t*).

(*c.*) By means of a perfectly clean pipette, half fill the
remainder of the tube with a *saturated* neutral solution of
zinc sulphate. Make *two* such electrodes.

(*d.*) Into each tube introduce a well amalgamated short piece of zinc wire with a thin copper wire soldered to its upper end (Z, Z), fix the electrodes in suitable holders in a moist chamber, and attach the wires of the zincs to the binding screws in the plate or stage of the moist chamber.

3. **The Brush-Electrodes** of v. Fleischl are very convenient. They consist of glass tubes 5 mm. in diameter and 4 cm. long. Into one end is fitted a perfectly clean camel's-hair pencil, and into the other dips a well-amalgamated rod of zinc with a binding screw at its free end. Place some clay in the lower part of the tube, and then fill it with a saturated solution of zinc sulphate. A piece of india-rubber tubing fits as a cap over the upper end of the glass tube. The brushes are moistened with a mixture of kaolin and normal saline.

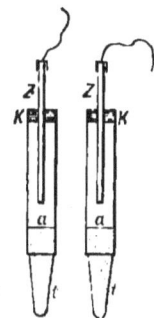

Fig. 80.—Non-polarisable Electrodes.— Z, Zincs; K, cork; *a*, zinc sulphate solution; *t, t*, clay points.

4. **The Shunt.**—This is an arrangement by which a greater or less proportion of a current can be sent through the galvanometer (Fig. 81).

The brass bars on the upper surface are marked with the numbers $\frac{1}{9}$, $\frac{1}{99}$, $\frac{1}{999}$, indicating the ratio between their resistance and that of the galvanometer, so that when the plug is inserted in positions afterwards to be mentioned, there may be $\frac{1}{10}$, $\frac{1}{100}$, or $\frac{1}{1000}$ of the whole current sent through the galvanometer.

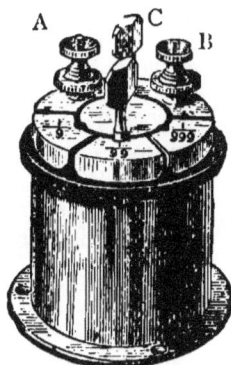

5. **Demarcation Current of Muscle.**

(*a.*) Arrange the apparatus according to the scheme (Fig. 82).

(*b.*) Introduce a shunt between the N.P. electrodes and the galvanometer.

Fig. 81.—The Shunt.

Connect two wires from the electrodes to the binding screws (A and B) of the shunt, and from the same binding screws attach two wires of the same kind to the galvanometer. Insert a plug (C) between A and B, whereby the muscle current is short-circuited. When working with muscle, keep a plug in the hole opposite $\frac{1}{6}$ on the shunt. Arrange the lamp and scale so as to have a good image of the mirror on the zero of the scale ; adjusting, if necessary, by means of the magnet moved by the milled head on the top of the glass shade (Fig. 82, m).

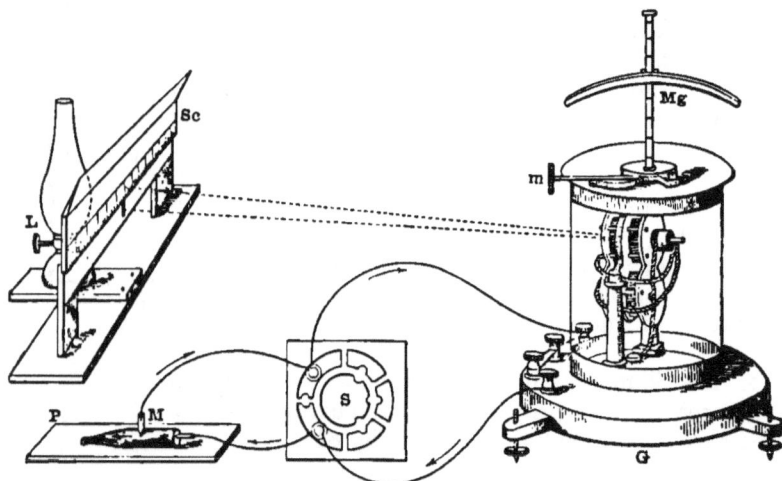

Fig. 82.—Arrangement of Apparatus for the Demarcation Current of Muscle.—M, Muscle on a glass plate, P ; S, shunt ; G, galvanometer ; Mg, its magnet moved by the milled head, m ; L and Sc, lamp and scale.

(c.) Test the electrodes, either by bringing them together, or by joining them with a piece of silk thread covered with china-clay paste. After removing all the plugs from the shunt, there ought to be no deflection of the spot of light. If there is none, there is no polarity, and the electrodes are perfect.

(d.) **Ascertain the Direction of Current in Galvanometer.**— Make a small Smee's battery with a two-ounce bottle. Place in the bottle dilute sulphuric acid (1 : 20) and two wires of zinc (–) and copper (+), with wires soldered to them. Connect them with the galvanometer. Arrange the shunt so that $\frac{1}{100}$ or $\frac{1}{1000}$ part of the current thus generated goes through the galvanometer. Note the deflection and its direction. Arrange the N.P. electrodes in the same way, and observe which is the negative and which the positive pole corresponding to the zinc and copper of the battery.

(*e.*) **Prepare a Muscle.**—Dissect out either the sartorius or semi-membranosus of a frog; these are selected because they consist of parallel fibres, but avoid touching the muscle with the acid skin of the frog. Lay the muscle on a glass plate or block of paraffin under the moist chamber with the N.P. electrodes.

(*f.*) Keep one plug in the shunt at C, so as to short-circuit the electrodes, and the other plug at $\frac{1}{0}$. Cut a fresh transverse section at one end of the muscle, and adjust the point of one electrode exactly over the centre (equator) of the longitudinal surface of the muscle. Apply the other electrode exactly to the centre of the freshly divided transverse surface (Fig. 82).

(*g.*) Remove the short-circuiting plug C from the shunt, keep one plug in at $\frac{1}{0}$, so that $\frac{1}{10}$ of the total current from the muscle goes through the galvanometer. Note the direction and extent of the deflection. By noting the direction, and from the observation already made (*d.*), one knows that the longitudinal surface of the muscle is $+$, and the transverse section $-$. Replace the plug key (C), and allow the needle to come to rest at zero.

(*h.*) Bring the N.P. electrode on the longitudinal surface nearer to the end of the muscle, and note the diminution of the deflection of the needle. Replace plug C.

(*i.*) Vary the position of the electrodes and note the variation in the deflection. If they be equi-distant from the equator, there is no deflection. The greatest deflection takes place when one electrode is over the equator and the other over the centre of the transverse section of a muscle composed of parallel fibres. The deflection, *i.e*, the electromotive force, diminishes as the electrodes are moved from the equator or the centre of the tranverse section. In certain positions no deflection is obtained.

6. Action Current of Muscle.

(*a.*) Use the same muscle preparation, or isolate the gastrocnemius with the sciatic nerve attached. Divide the muscle transversely, and lay the artificial transverse section on one electrode, and the longitudinal surface on the other. Observe the extent of the deflection.

(*b.*) Adjust a du Bois induction apparatus for interrupted shocks, placing it at some distance from the galvanometer.

(*c.*) Take the demarcation current, observing the deflection, and allow the spot of light to take up its new position on the scale. Stimulate the muscle by the ordinary electrodes to throw it into tetanus, and observe that the spot of light travels towards zero. This was formerly called the " **negative variation of the muscle current.**" It is now called the " **action current** " of muscle. If the gastrocnemius be used, stimulate the sciatic nerve. Care must be taken that the muscle does not shift its position on the electrodes.

LESSON XXXIX.

DEMARCATION AND ACTION CURRENT IN NERVE—ELECTRO-MOTIVE PHENOMENA OF THE HEART—CAPILLARY ELECTROMETER.

1. Demarcation Current of Nerve.

(*a.*) Render the galvanometer as sensitive as possible by adjusting at a suitable height the north pole of the magnet over the north pole of the upper needle.

(*b.*) Prepare N.P. electrodes for a nerve.   In this case,

the electrodes are hook-shaped, and one is adjusted over the other. The upper hooked electrode has a groove on its concavity communicating with the interior of the tube (Fig. 83). Place only one plug in the shunt between A and B.

(*c.*) Dissect out a long stretch of the sciatic nerve, make a fresh transverse section at either end, hang it over the upper N.P. electrode (N), and resting with its two cut ends on the lower electrode (C), thus doubling the strength of the current (Fig. 83).

(*d.*) Remove the plug from C in the shunt, and pass the whole of the demarcation-nerve current through the galvanometer, noting the deflection.

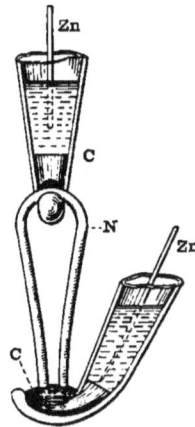

Fig. 83.—Nerve N. P. Electrodes.— N, Nerve; C, clay of electrodes; Zn, Zincs.

(*e.*) Instead of adjusting the nerve as in (*c.*), it may be so placed on the ordinary tube N.P. electrodes, that the cut end rests on one electrode, and the longitudinal surface on the other, thus leaving part of the nerve free. Observe the deflection in this way.

2. Action Current of Nerve.

(*a.*) Observe the amount of deflection as in (*e.*) Stimulate with an interrupted current the free end of the nerve, and observe that the spot of light travels towards zero. This was formerly called the "negative variation" of the nerve current.

3. Electro-motive Phenomena of the Heart.—The arrangement of the apparatus is precisely the same as in Lesson XXXVIII.

(*a.*) Make a Stannius preparation of the heart, using only the first ligature (Lesson XLVI., 1) to arrest the heart's action. Lead off with brush N.P. electrodes from base and apex of the heart ; there is no deflection.

(*b.*) Pinch the apex so as to injure it, it becomes negative.

13

(c.) Excise a heart so as to get a spontaneously beating ventricle, lead off from the base and apex of the latter, observe the so-called "negative variation" with each contraction.

4. Capillary Electrometer.

(a.) Lead off a muscle to the two binding screws of a capillary electrometer. The fine thread of mercury must be observed with a microscope.

LESSON XL.

GALVANI'S EXPERIMENTS — SECONDARY CONTRACTION AND TETANUS— PARADOXICAL CONTRACTION—KÜHNE'S EXPERIMENT.

1. Galvani's Experiments.

(a.) Destroy the brain of a frog, divide the spine about the middle of the dorsal region, cut away the upper part of the body, and remove all the viscera. Remove the skin from the hind legs, divide the iliac bones and urostyle, taking care to avoid injuring the lumbar plexus, which will remain as the only tissue connecting the lower end of the vertebral column with the legs. Thrust an S-shaped copper hook through the lower end of the spine and spinal cord (Fig. 84).

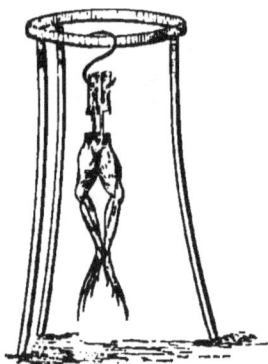

Fig. 84.—Galvani's Experiment.

(b.) Hang the frog by means of the hook to an ordinary iron tripod. Tilt the tripod so that the legs of the frog come in contact with one of the legs of the tripod; vigorous contractions occur whenever the frog's legs touch the tripod.

(c.) Allow the frog to hang perpendicularly without touching the tripod, and take a U-shaped piece of wire composed of a copper and zinc wire soldered together. Touch the nerves above with the copper (or zinc) end, and the muscles below with the zinc (or copper), when contraction occurs at make, or break, or both.

2. Contraction without Metals.—Take a strong frog, make a nerve-muscle preparation, leaving the leg attached to the femur, and having the sciatic nerve as long as possible. Hold the femur in one hand, lift the nerve on a camel's-hair pencil moistened with normal saline, and allow it to fall upon the gastrocnemius, when the muscle will contract. Contraction occurs because the nerve is suddenly stimulated, owing to the surface of the muscle having different potentials.

3. Secondary Contraction and Secondary Tetanus.

(*a.*) Arrange the induction apparatus for single make and break shocks. Pith a frog, dissect out two nerve-muscle preparations, as in Fig. 85.

(*b.*) Place the left sciatic nerve (A) over the right gastrocnemius (B), or thigh muscles, and the right sciatic nerve over the electrodes (E).

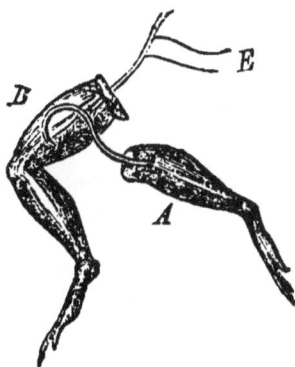

(*c.*) Stimulate the nerve with single induction shocks, and note that the muscles of both B and A contract. The contraction in A is called a **secondary contraction.**

(*d.*) Arrange the induction machine for interrupted shocks, and stimulate the nerve at E, B is thrown into tetanus, and so is A simultaneously. This is **secondary tetanus.** This is a

Fig. 85.—Secondary Contraction.

proof of the "action current" in muscle. The nerve of A is stimulated by the variation of the muscle current during the contraction of B.

(*e.*) Ligature the nerve of A near the muscle, stimulate the nerve of B; there should be no contraction of A although B contracts.

(*f.*) Prepare another limb and adjust it in place of A, ligature the nerve of B. On stimulating the nerve of B, no contraction takes place either in A or B.

4. Secondary Contraction.

(*a.*) Make a nerve-muscle preparation, and place it on a glass plate (B). Dissect out a long stretch of the sciatic nerve of the opposite side (A). Lay 1 cm. of the isolated sciatic nerve (A) on a similar length of the nerve of the nerve-muscle preparation (B), (Fig. 86).

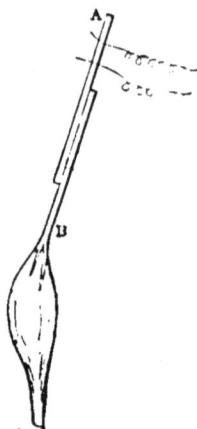

Fig. 86.—Scheme of Secondary Contraction.

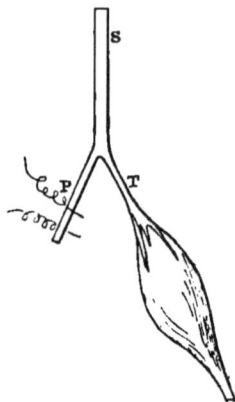

Fig. 87.—Scheme of Paradoxical Contraction.

(*b.*) Stimulate A with a single induction shock, the muscle of B contracts. Stimulate A with an interrupted current, the muscle of B is thrown into tetanus.

(*c.*) Ligature A and stimulate again, B does not contract. Therefore, its contraction was not due to an escape of the stimulating current. The "secondary contractions" in B are due to the sudden variations of the electromotivity produced in A when it is stimulated.

5. Paradoxical Contraction.

(*a.*) Arrangement.—Arrange a Daniell's cell and key for giving a galvanic current.

(*b.*) Pith a frog, expose the sciatic nerve down to the knee (Fig. 87, S). Trace the two branches into which it divides. Divide one of the branches as near as possible to the knee, and stimulate its *central end* (P). The muscles supplied by the other division of the nerve (T) contract.

6. Kühne's Experiment.

(*a.*) Invert an ordinary earthenware bowl (B), and fix with wax to its base a piece of glass 10 cm. square (Fig. 88, G).

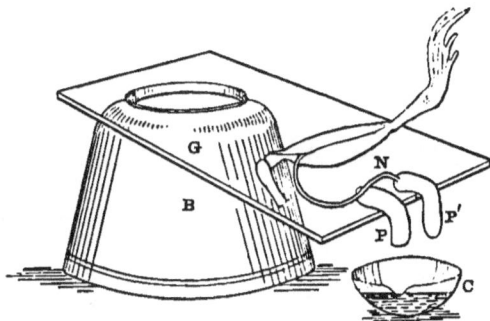

Fig. 88.—Kühne's Experiment.—B, Bowl ; G, glass plate ; N, nerve on P, P', pads of clay ; C, capsule.

(*b.*) Roll two plugs of kaolin (moistened with normal saline), about 1 cm. in diameter and 6 cm. in length (P, P'), bend them at a right angle, and hang them over the glass plate about 6 mm. apart.

(*c.*) Make a nerve-muscle preparation, lay the muscle on the glass plate, and the nerve over the two rolls of china clay (Fig. 88, N).

(*d.*) Fill a small glass vessel (C) with normal saline, and allow the two free ends of the clay to dip into it. With each dip the muscle contracts. In this case the nerve is stimulated by the completion of the circuit of its own demarcation current.

LESSON XLI.

ELECTROTONUS—ELECTROTONIC VARIATION OF THE EXCITABILITY.

1. Electrotonus.—When a nerve is traversed by a constant current, its vital properties are altered—*i.e.*, its **excitability, conductivity,** and **electromotivity.** The region of the nerve affected by the positive pole is said to be in the **anelectrotonic,** and that by the negative in the **kathelectrotonic** condition. Therefore we have to study the—

I.—Electro-motive alteration of the *excitability and conductivity.*

II.—Electro-motive alteration of the *electromotivity.*

2. Electrotonic Variation of the Excitability—Apparatus required.—Three Daniell's cells, two pairs of N.P. electrodes, two du Bois keys, a Morse or spring key, commutator with cross-bars, or Thomson's reverser, induction machine, wires, moist chamber, drum, frogs, and the usual instruments.

A. (*a.*) Arrange the apparatus according to the scheme (Fig. 89) introducing the Morse key in the primary circuit. Prepare two pairs of N.P. electrodes for the nerve.

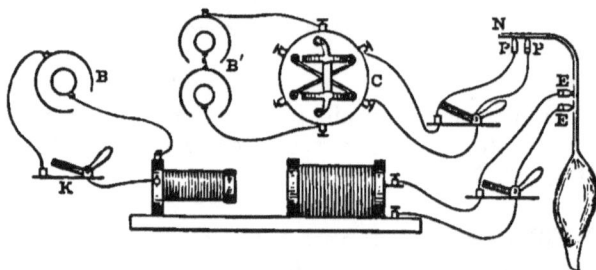

Fig. 89.—Scheme of Electrotonic Variation of Excitability.—P P, Polarising, and E E, stimulating current.

(*b.*) Take two Daniell's cells and connect them with a Pohl's commutator with cross-bars (C), or Thomson's reverser; connect the commutator—a short-circuiting key intervening—to one pair of the N.P. electrodes. Call this the "polarising current" (P P).

(c.) Arrange an induction machine for single shocks, the other pair of N.P. electrodes being in the secondary circuit, arranged for short-circuiting. Call this the "exciting current" (E E).

(d.) Make a nerve-muscle preparation with the nerve as long as possible, and arrange it for recording its movements on a drum. Arrange the nerve on the two pairs of electrodes in the moist chamber, the "polarising" pair being next the cut end of the nerve (P P), and about 1 centimetre apart. Between the polarising pair and the muscle, apply the "exciting" pair of electrodes to the nerve (E E).

(e.) With the polarising current short-circuited, pull away the secondary from the primary coil, and find the minimum distance at which a feeble contraction of the muscle is obtained. Push the secondary coil up a little to obtain just a weak contraction, and take a tracing. Previously arrange the commutator to send a *descending* current through the nerve. While the muscle is contracting feebly throw in the descending polarising current, at once the contraction becomes much *stronger*. Reverse the commutator to send an *ascending* polarising current through the nerve, and the contraction will cease.

(f.) Repeat the experiment using Neef's hammer.

In the first case, the area stimulated by the exciting electrodes was affected by the negative pole—*i.e.*, was in the condition of kathelectrotonus, and the tetanus was increased; therefore, *the kathelectrotonic condition increases the excitability of a nerve.* In the second case, the nerve next the exciting electrodes was in the condition of anelectrotonus, and the contractions ceased; therefore, *the anelectrotonic condition diminishes the excitability of a nerve* (Fig. 90).

B. Another Method.—(a.) Connect two small Grove's cells or two Daniell's to a Pohl's commutator *with cross-bars*, introducing a du Bois key to short-circuit the battery. Fill the cups with mercury, and place the commutator in a

small tray to avoid spilling the mercury. From two of
the binding screws connect wires with two N.P. electrodes

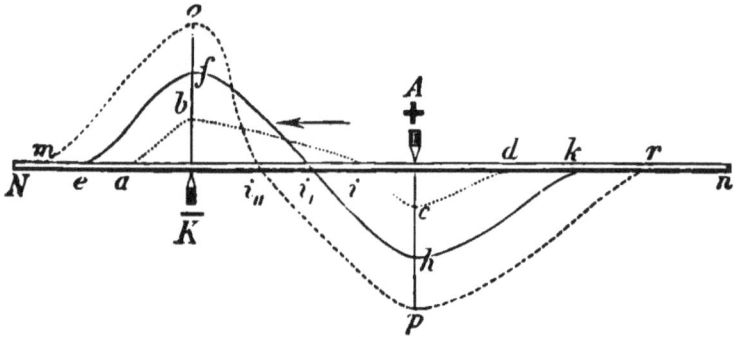

Fig. 90.—Scheme of Electrotonic Variation of Excitability in a Nerve.—
K, Cathode; A, anode; N n, nerve. The curve above the line indicates increase, and that below the line decrease, of excitability.

or the platinum electrodes of du Bois, introducing a short-
circuiting key in the electrode circuit (Fig. 91).

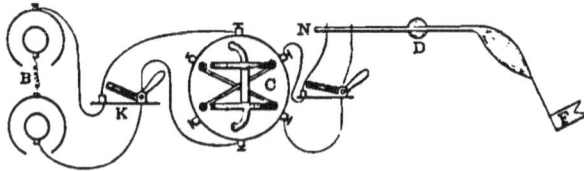

Fig. 91.—Scheme of Electrotonic Variation of Excitability.—D, Drop of
strong solution of salt on the nerve, N ; F, flag on the muscle.

(*b.*) Make a nerve-muscle preparation, attach a straw flag
to the foot, as directed in Lesson XXVII., 4, and clamp the
femur in a clamp on a stand, as in Fig. 53. Lay the nerve
over the electrodes. Trace the direction of the current,
and make a mark to guide you as to when the current in
the nerve is descending or ascending—*i.e.*, whether the
negative or positive pole is next the muscle.

(*c.*) Place a drop of a saturated solution of common salt
on the nerve between the electrodes and the muscle. In a
minute or less the toes begin to twitch, and by-and-by all the
muscles of the leg become tetanic, so that the flag is raised
and kept in the horizontal position. .

(*d.*) Turn the commutator so that the positive pole is next the muscle, at once the straw falls—*i.e*, the excitability of the nerve in the region of the positive pole is so diminished as to "block" the impulse passing to the muscle, showing that the positive pole lowers the excitability.

(*e.*) Reverse the commutator so that the negative pole is next the muscle. At once the limb becomes tetanic, the negative pole (*kathelectrotonic area*) increases the excitability.

LESSON XLII.

ELECTROTONIC VARIATION OF THE ELECTROMOTIVITY — PFLÜGER'S LAW OF CONTRACTION—RITTER'S TETANUS.

1. The Electrotonic Variation of the Electromotivity.

(*a.*) Arrange a long nerve on the N.P. electrodes, as for determining its demarcation current. Place the free end of the nerve on a pair of N.P. electrodes—the polarising current—arranged as in Lesson XLI., 2, **A.,** so that the current can be made ascending or descending.

(*b.*) Take the deflection of the galvanometer needle or demarcation current when the polarising current is shut off. Throw in a descending polarising current, and observe that the spot of light travels towards zero. Reverse the commutator, and throw in an ascending current, the spot of light shows a greater positive variation than before. From this we conclude that *kathelectrotonus diminishes the electromotivity, while anelectrotonus increases it.*

2. Pflüger's Law of Contraction—Apparatus.—Several small Grove's cells, commutator with cross-bars, du Bois and Morse key, rheochord, N.P. electrodes, moist chamber, wires, frog, recording apparatus, and usual instruments.

(*a.*) Arrange the apparatus as in the scheme (Fig. 92). Take two Daniell or small Grove cells, connect them to a Pohl's commutator with cross-bars, and introduce a Morse or mercury key (K) into the circuit; connect the commutator with the rheochord (R). Connect the rheochord with a pair

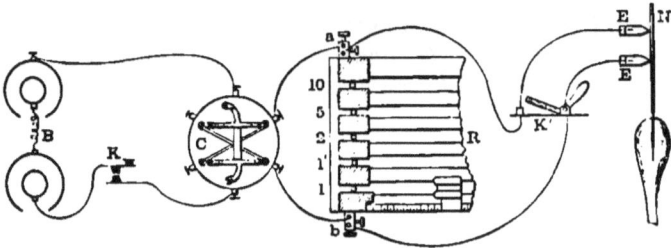

Fig. 92.—Scheme for Pflüger's Law.—R, Rheochord.

of N.P. electrodes with a short-circuiting key introduced. Fix to a lever a nerve-muscle preparation—with a long nerve—in the moist chamber, and lay the nerve over the electrodes.

(*b.*) Begin with all the plugs in position in the rheochord and the slider hard up to the brass blocks. Place the commutator to give an ascending current, make and break the current—gradually adjusting the slider—until a contraction occurs at make and none at break. Reverse the commutator to get a descending current, make and break, observing again a contraction at make and none at break. This represents the effect of a *weak* current.

(*c.*) Pull the slider further away and remove one or more plugs, until contraction is obtained at make and break, both with an ascending and descending current. This represents the effect of a *medium* current.

(*d.*) Use six small Grove's cells, take out all the plugs from the rheochord, and with the current ascending, contraction occurs at break only—while with a descending current, contraction occurs only at make. This represents the effect of a *strong* current. Tabulate the results in each case.

For this experiment very fresh and strong frogs are necessary, and several preparations will be required to work out all the details of the law. Instead of reversing the commutator after testing the effect of an alteration of the direction of the current, the student may use one preparation to test at intervals the effect of weak, medium, and strong currents, when the current is ascending, and a second preparation to test the results with currents of varying intensity when the current is descending. The results may be tabulated as follows :—R = rest ; C = contraction.

STRENGTH OF CURRENT.	ASCENDING.		DESCENDING.	
	On Making.	On Breaking.	On Making.	On Breaking.
Weak, . .	C	R	C	R
Medium, .	C	C	C	C
Strong, . .	R	C	C	R

3. Ritter's Tetanus.

(*a.*) Connect three Daniell's cells with non-polarisable electrodes short-circuiting with a du Bois key. Prepare a nerve-muscle preparation, and apply the electrode to the nerve so that the + pole is next the muscle—*i.e.*, the current is ascending in the nerve. Allow the current to circulate in the nerve for some time (usually about five minutes is sufficient), no contraction takes place. Short-circuit, and observe that the muscle becomes tetanic.

(*b.*) Divide the nerve between the electrodes, and the tetanus does not cease ; but on dividing it between the + pole and the muscle, the tetanus ceases. Therefore the tetanus is due to some condition at the positive pole.

LESSON XLIII.

VELOCITY OF NERVE ENERGY IN A MOTOR NERVE—DOUBLE CONDUCTION IN NERVE —KÜHNE'S GRACILIS EXPERIMENT.

1. Velocity of Nerve Energy in a Motor Nerve.

The rate of propagation of a nerve impulse may be estimated by either the pendulum or spring myograph. With slight modifications the two processes are identical, only in using the spring myograph it is necessary to use such a coiled spring, as will cause the glass plate to move with sufficient rapidity, to give an interval long enough for the easy estimation of the latent period.

(*a.*) Use the **spring myograph** and arrange the experiment according to the scheme (Fig. 93)—*i.e.*, an induction coil for

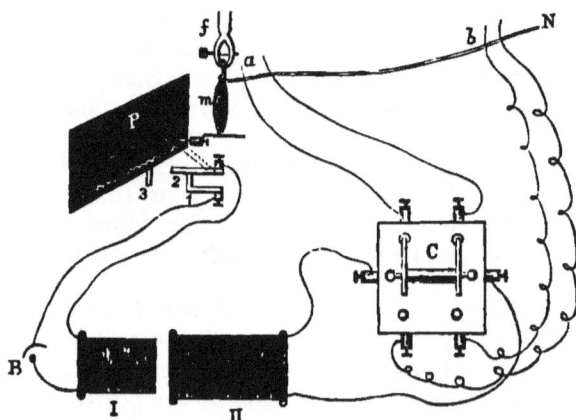

Fig. 93.—Scheme for Estimating the Velocity of Nerve Energy.

single shocks with the trigger-key of the myograph (1, 2) arranged in the primary circuit; in the secondary circuit (which should be short-circuited, not represented in the diagram) place a Pohl's commutator *without cross-bars* (C).

Two pairs of wires from the commutator pass to two pairs of electrodes (a, b), movable on a bar within the moist chamber. Measure the distance between the electrodes.

(b.) Make a nerve-muscle preparation with the nerve as long as possible (N), clamp the femur (f), attach the tendon (m) to a writing-lever, and lay the nerve over the electrodes, the distance between them being known.

(c.) Arrange the glass plate covered with smoked paper, adjust the lever to mark on the glass, close the trigger-key in the primary circuit, and un-short-circuit the secondary. Turn the bridge of the commutator so that the stimulus will be sent through the electrodes next the muscle (a). Press the thumb-plate, and shoot the glass plate. The tooth (3) breaks the primary circuit, and a curve is inscribed on the plate.

(d.) Short-circuit again, replace the glass plate, close the trigger-key, reverse the commutator. This time the stimulus will pass through the electrodes away from the muscle (b). Un-short-circuit the secondary circuit, and shoot the glass plate. Again another curve will be inscribed, this time a *little later* than the first one.

(e.) Replace the glass plate, close the trigger-key, short-circuit the secondary circuit, and shoot the plate. This makes the abscissa.

(f.) Replace the glass plate, close the trigger-key, and bring the tooth of the glass plate (3) just to touch the trigger-key; raise the writing-lever to make a vertical mark. This indicates the moment when the stimulus was thrown into both points of the nerve.

(g.) Remove the moist chamber, push up the glass plate, close the trigger-key, and arrange a tuning-fork vibrating 250 D.V. per second to write under the abscissa. Shoot the plate again and the time-curve will be obtained. Fix the tracing, draw ordinates from the beginning of the curves obtained by the stimulation of a and b respectively, measure the time between them from the time-curve (this gives the time the impulse took to travel from b to a), and calculate the velocity from the data obtained.

Example.—Suppose the length of nerve to be 4 cm., and the time required for the impulse to travel from b to a to be $\frac{1}{750}$ sec. Then we have $4 : 100 : \frac{1}{750} : \frac{1}{30}''$, or 30 metres (about 90 feet) per second, as the velocity of nerve energy along a nerve.

2. Repeat the observation with the **pendulum myograph**. Practically the same arrangements are necessary.

If it be desired to test the effect of heat or cold on the rapidity of propagation, the nerve must be laid on ebonite electrodes, made in the form of a chamber, and covered with a lacquered copper plate on which the nerve rests. Through the chamber water at different temperatures can be passed, and the effect on the rate of propagation observed.

3. **Unequal Excitability of a Nerve—Apparatus.**—Battery, two keys, wires, commutator, induction machine, two pairs of electrodes.

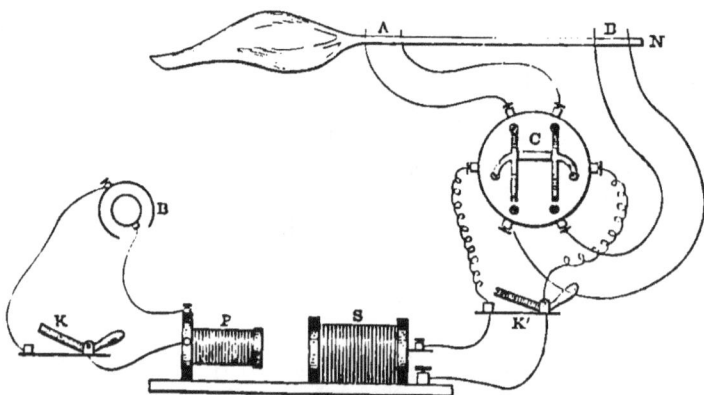

Fig. 94.—Scheme for the Unequal Excitability of a Nerve.

(*a.*) Arrange the apparatus as in Fig. 94, introducing a Morse key in the primary circuit. Dissect out the whole length of the sciatic nerve with the foot attached. Lay the nerve on two pairs of electrodes, A and B, one near the muscle, and the other away from it, and as far apart as possible. Two pairs of wires thrust through a cork or piece of caoutchouc will do quite well.

(*b.*) Stimulate the nerve at A with a strength of current that gives just a minimal contraction. Reverse the commutator, and on stimulating at B a much stronger contraction is obtained, because the excitability of a nerve is greater further from a muscle.

4. Double Conduction in Nerve—Kühne's Experiment on the Gracilis.—The gracilis is divided into a larger and smaller portion (L) by a tendinous inscription (K) running across it (Fig. 95). The nerve (N) enters at the hilum in the larger half, and bifurcates, giving a branch (k) to the smaller portion, and another to the larger portion of the muscle.

(*a.*) Excise the gracilis from a large frog, and cut it as shown in Fig. 96, avoiding injury to the nerves, so that only the nerve twig (k) connects the larger and smaller halves, and in one tongue (Z) terminates a nerve. The gracilis after excision must be laid on a glass plate with a *black* background, else one does not see clearly the inscription and the course of the nerves. Both are easily seen on the black surface.

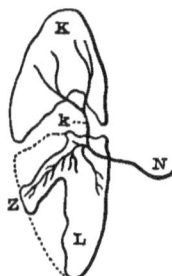

(*b.*) Stimulate the tongue (Z) with fine electrodes about 1 mm. apart, and contraction occurs in both L and K. This can be due only to centripetal conduction in a motor nerve, and this experiment is adduced by Kühne as the best proof of double conduction in nerve fibres.

Fig. 95.—Kühne's Experiment on the Gracilis.

PHYSIOLOGY OF THE CIRCULATION.

LESSON XLIV.

THE FROG'S HEART—BEATING OF THE HEART—EFFECT OF HEAT AND COLD —SECTION OF THE HEART.

1. The Heart of the Frog and how to expose it.

(*a.*) Pith a frog, and lay it on its back on a frog-plate. Make a median incision through the skin over the sternum, and from the middle of this make transverse incisions.

(*b.*) Reflect the four flaps of skin, raise the lower end of the sternum with a pair of forceps, and cut through the sternal cartilage just above its lower end to avoid wounding the epigastric vein. With a strong pair of scissors cut along the margins of the sternum, and divide it above transversely to remove the anterior wall of the thorax. This exposes the heart, still enclosed within its pericardium, where it can be seen beating.

(*c.*) With a fine pair of forceps carefully lift up the thin transparent pericardium, cut it open, thus exposing the heart.

2. Study the General Arrangement of the Frog's Heart.

(*a.*) Observe its shape, noting the two auricles above (Ad, As), and the conical apex of the single ventricle below (v), the auricles being mapped off from the ventricle by a groove ·or furrow which runs obliquely across its anterior aspect. The ventricle is continuous anteriorly with the bulbus aortæ (B), which projects in front of the right auricle, and divides into two aortæ—right and left, the left being the larger (Fig. 96).

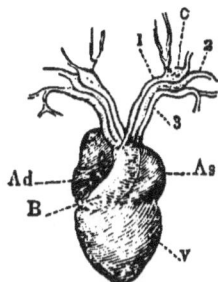

Fig. 96.—Frog's Heart from the front.—V, Single ventricle ; Ad, As, right and left auricles ; B, bulbus arteriosus ; 1, carotid ; 2, aorta ; 3, pulmocutaneous arteries ; C, carotid gland.

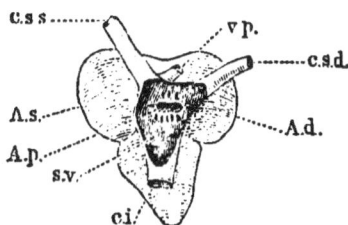

Fig. 97.—Heart of Frog from behind.—s.v, Sinus venosus opened ; c.i, inferior; c.s.d, c.s.s, right and left superior venæ cavæ ; v.p, pulmonary vein ; Ad and As, right and left auricles ; Ap, communication between the right and left auricle.

(b.) Tilt up the ventricle and observe the sinus venosus (Fig. 97, s.v), continuous with the right auricle, and formed by the junction of the large inferior vena cava (c.i), and the two (smaller) superior venæ cavæ (c.s.s, c.s.d).

3. The Heart beats after it is excised.

(a.) With a seeker tilt up the apex of the ventricle, and observe that a thin thread of connective-tissue, called the "frænum," containing a small vein, passes from the pericardium to the posterior aspect of the ventricle. Divide it with a fine pair of scissors. Count the number of beats per minute. Seize with forceps the part of the frænum attached to the ventricle, and lift up the heart therewith ; and with a sharp pair of scissors cut out the heart by dividing the inferior vena cava, the two superior venæ cavæ, and the two aortæ. Place the excised heart in a watch-glass, and cover it with another watch-glass.

(b.) The heart goes on beating. Count the number of beats per minute. Therefore its beat is **automatic**, and the heart contains within itself the mechanism for originating

14

and keeping up its rhythmical beats. If the heart tends to become dry, moisten it with normal saline solution, although normal saline containing a little blood is better.

(c.) Observe also how during diastole the heart is soft and flaccid, and takes the shape of any surface it may rest on, while during systole when it contracts, it becomes harder, while the apex is raised and erected.

4. Effect of Heat and Cold on the Excised Heart.

(a.) Place the watch-glass containing the beating heart on the palm of the hand, and the heart will beat faster; or place it over a beaker containing warm water, which must not be above 40°C. Observe that as the temperature rises, the heart beats faster—i.e., there are more beats per minute, also that each single beat is faster.

(b.) Remove the watch-glass from the palm, place it over a beaker containing cold water or ice, when the number of beats will diminish, each beat being executed more slowly and sluggishly.

5. Section of the Heart.

(a.) With a sharp pair of scissors divide the ventricle at its upper third just below the auriculo-ventricular groove. Observe that the auricles with the upper third of the ventricle attached to them continue to beat spontaneously, while the lower two-thirds of the ventricle no longer beat spontaneously. If it be pricked with a needle, however, it contracts just as often as it is stimulated mechanically. It responds by a single contraction to a single stimulus, but a single stimulus does not excite a series of rhythmical contractions.

(b.) With a sharp pair of scissors divide the auricles with the attached portion of the ventricle longitudinally. Each half continues to contract spontaneously, although it is possible that the rhythm may not be the same in both.

6. Movements of the Heart.—Expose the heart of a freshly pithed frog as directed in Lesson XLIV., 1, or better still,

destroy only the brain and then curarise the frog. After dividing the pericardium and exposing the heart, observe

(*a.*) That the two auricles contract synchronously and force their blood into the ventricle, which from being pale and flaccid becomes red, turgid, and distended with blood.

(*b.*) That immediately the ventricle suddenly contracts, and forces the blood into the bulbus aortæ, at the same time becoming pale, while its apex is tilted forwards and upwards. As the auricles continue to fill during the systole of the ventricle, on superficial observation it might seem as if the blood were driven to and fro between the auricles and ventricle, but careful observation will soon satisfy one that this is not the case. Observe very carefully how the position of the auriculo-ventricular groove varies during the several phases of cardiac activity.

(*c.*) The slight contraction of the bulbus aortæ immediately following the ventricular systole.

(*d.*) The diastolic phase or pause when the whole heart is at rest before the auricles begin to contract.

(*e.*) Ligature the frænum and divide it, tilt up the ventricle by the ligature attached to the frænum, and observe the sinus venosus. The peristaltic wave, or wave of contraction, begins at the upper end of the vena cava inferior and sinus venosus; it extends to the auricles, which contract, then follows the ventricular systole and that of the bulbus aortæ, and finally, the pause, when the whole sequence of events begins again with the systole of the sinus.

(*f.*) Before the ventricular systole is complete the sinus is full, while the auricles are filling.

LESSON XLV.

GRAPHIC RECORD OF THE FROG'S HEART — EFFECT OF TEMPERATURE.

1. Graphic Record of the Contracting Frog's Heart.

(*a.*) Pith a frog, or destroy its brain, and then curarise it. Expose the heart, still within its pericardium, and arrange a heart-lever, so that it rests lightly on the pericardium over the beating heart. Adjust the lever to write on a revolving cylinder, moving at a suitable rate (5–6 cm. per second). Take a tracing of the beating of the heart.

(*b.*) A suitable **heart-lever** is easily made with a straw about 12 inches long, or a thin strip of wood about the same length. Thrust a needle transversely either through the straw or wood, or through a piece of cork slipped over the straw about 2 inches from one end of the lever. The needle forms the fulcrum of the lever, and works in bearings, whose height can be adjusted. To the end of the lever nearest this is attached at right angles a needle with a small piece of cork on its free end. The lever is so adjusted that the cork on the needle rests on the heart. The long arm of the lever is provided with a writing-style of copperfoil, or a writing point made of parchment paper, fixed to it with sealing-wax. By using a long lever a sufficient excursion is obtained.

(*c.*) Open the pericardium, expose the heart, and adjust the cork on the lever. To obtain a good tracing, it is well to put some resistant body behind the heart. Raise up the ventricle, ligature the frænum, and divide the latter outside the ligature, and behind the heart place a pad of blotting-paper moistened with normal saline, or a thin cover slip. Adjust the lever, with its cork pad, on the junction of the auricles and ventricle, to write on the cylinder, moving at a slow rate (5–6 cm. per second), and take a tracing, noting the rise and fall of the lever.

(*d.*) Fix the tracings, and observe in the tracing a first ascent due to the auricular contraction, and succeeding this a second ascent due to the contraction of the ventricle, followed by a slow subsidence due to the continuation of the ventricular systole, and then a sudden descent due to the diastolic relaxation of the heart.

2. **Auricular Contraction.**—Adjust the lever again so that it rests on the auricles alone, and take a tracing. Note the smaller excursion of the lever. In this case the cork resting on the auricles must be small.

3. **Ventricular Contraction.**—Adjust the lever so as to obtain a tracing of the ventricular movements only.

4. In all the above experiments arrange an electro-magnetic time-marker (Fig. 67) under the recording lever, so that the points of the recording lever and time-marker write exactly in the same vertical line with each other. In this way one can calculate the time-relations of any part of the curve. The time-marker is arranged to record seconds, and is driven by an electric clock.

5. **Effect of Varying Temperatures on the Excised Heart.**

(*a.*) Excise the heart of a pithed frog, lay it on a cylindrical brass cooling-box, three inches long and one broad, fixed to a support, and fitted with an inlet and outlet tube, like that in Fig. 62. Fix india-rubber tubes to the inlet and outlet tubes of the cooling-box, the inlet tube passing from a funnel fixed in a stand above the box, and the outlet tube discharging into a vessel below it. Adjust the heart lever to record the movements of the contracting ventricle on a slowly revolving drum. If the heart tends to become dry, moisten it with normal saline mixed with blood. Adjust a time-marker, as indicated for other experiments. Take a tracing.

(*b.*) Pass water from 10° to 20° C. through the cooling-box, noting the effect on the number of the contractions, and the duration, height, and form of each single beat.

STANNIUS'S EXPERIMENT AND INTRA-CARDIAC INHIBITORY MOTOR CENTRE.

1. Stannius's Experiment.—Pith a frog, and expose its heart in the usual way.

(*a.*) With a seeker clear the two aortæ from the auricles, and with an aneurism needle pass a moist stout thread between the two aortæ and the superior venæ cavæ, turn up the apex of the heart, divide the frænum, and raise up the whole heart to expose its posterior surface, and the crescent or line of junction of the sinus venosus and the right auricle. Bring the two ends of the thread ligature round the heart—call this for convenience No. 1 ligature—tie them, and tighten the ligature just over the "crescent," so as to constrict the line of junction of the sinus venosus with the right auricle. Before tightening the ligature, observe that the heart is beating freely. On tightening the ligature, the auricles and ventricle cease to beat, and remain in a state of relaxation, while the sinus venosus continues to beat at the same rate as before. After a time, if left to itself, the ventricle begins to beat, but with an altered rhythm. If the relaxed ventricle be pricked, it executes a single contraction.

(*b.*) When the heart is still relaxed, take a second ligature (No. 2), and preferably of a different colour to distinguish it from No. 1, place it round the heart, and tighten it over the auriculo-ventricular groove, so as to separate the ventricle from the auricles. Immediately the ventricle begins to beat again, while the auricles remain relaxed or in diastole.

(*c.*) Instead of applying No. 2 ligature, the ventricle may be cut off from the auricles by means of a pair of scissors. Immediately after it is amputated, the ventricle begins to beat.

2. Intra-Cardiac Inhibitory Centre.

(*a.*) Expose the heart in a pithed frog, tie a fine silk ligature round the frænum, and divide the latter between the ligatured spot and the pericardium. Gently raise the whole heart upwards to expose the somewhat whitish **V**-shaped "crescent" between the sinus venosus and the right auricle.

(*b.*) Prepare previously an induction machine arranged to give an interrupted current. Place the electrodes—which must be fine, and their points not too far apart (2 millimetres)—upon the crescent, and faradise it for a second; if the current be sufficiently strong, the auricles and ventricle cease to beat for a time, but they begin to beat even in spite of continued stimulation.

(*c.*) Stimulate the auricles, there is no inhibition or arrest.

(*d.*) If a drop of solution of sulphate of atropia (Lesson XLVIII., 1) be applied to the heart, stimulation of the crescent no longer arrests the action of the heart, for the atropine paralyses the inhibitory fibres of the vagus.

3. Seat of the Motor Centres.

(*a.*) Expose a pithed frog's heart, cut out the ventricle with the auricles attached to it, and observe that the heart continues to beat. Divide the ventricle vertically by two parallel cuts into three portions. The middle portion contains the auricular septum, in which lie ganglionic cells. It continues to beat while the right and left lateral parts do not beat spontaneously, but respond by means of a single contraction if they are stimulated.

LESSON XLVII.

CARDIAC VAGUS AND SYMPATHETIC OF THE FROG AND THEIR STIMULATION.

1. Cardiac Vagus of the Frog—How to Expose it.—In this case a preliminary dissection must be made before the student attempts to stimulate the vagus.

(*a.*) Pith a frog, or destroy its brain and curarise it. Lay it on its back on a frog-plate. Expose the heart, remove the sternum, and pull the fore-legs well apart. Introduce a small test-tube, or stick of sealing-wax into the œsophagus, to distend it; the nerves leaving the cranium are better seen winding round from behind when the œsophagus is distended. Remove the muscles covering the petrohyoid muscles which reach from the petrous bone to the posterior horn of the hyoid bone (Fig. 98). Three nerves are seen coursing round the pharynx parallel to these muscles. The lowest is the hypoglossal (H), easily recognised by tracing it forward to the tongue, above it is the vagus in close relation

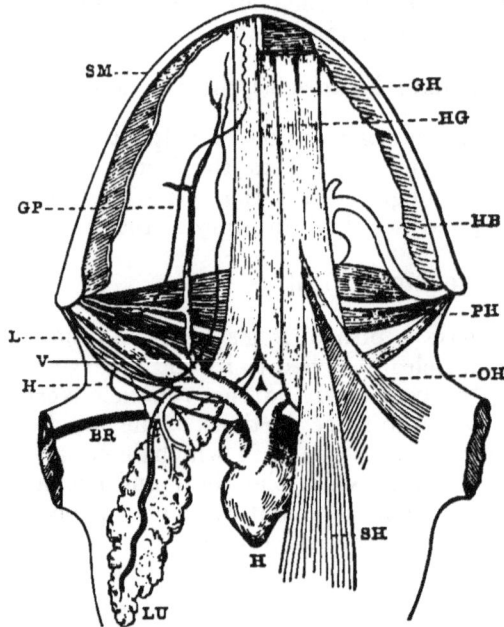

Fig. 98.—Scheme of the Dissection of the Frog's Vagus.—SM, Submentalis; LU, lung; V, vagus; GP, glossopharyngeal; H, hypoglossal; L, Laryngeal; PH, SH, GH, OH, petro-, sterno-, genio-, omo-hyoid; HB, hyoid; HG, hyoglossus; H, heart; BR, brachial plexus.

with a blood-vessel (V), and still further forward is the glosso-pharyngeal (GP). Observe the laryngeal branch of the vagus (L). The vagus, as here exposed outside the cranium, is really the vago-sympathetic. The glossopharyngeal and vagus leave the cranium through the same foramen in the ex-occipital bone, and through the same foramen the sympathetic enters the skull.

2. Stimulation of the Cardiac Vagus.

(*a.*) Adjust a heart-lever so as to record the contractions of the heart on a revolving drum moving at a very slow rate.

(*b.*) Place well-insulated electrodes under the trunk of the vagus, stimulate it with an interrupted current, and observe that the whole of the heart is arrested in diastole. Although the faradisation is continued the heart recommences beating. The arrest, or *period of inhibition*, is manifest in the curve by the lever recording merely a straight line. If

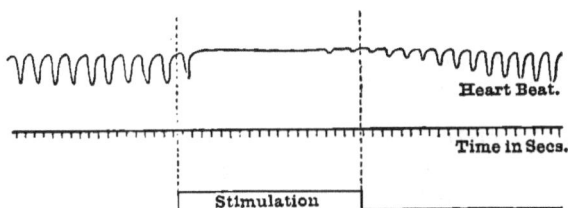

Fig. 99.—Vagus Curve of Frog's Heart.

the laryngeal muscles contract, and thereby affect the position of the heart, divide the laryngeal branch of the vagus.

(*c.*) Note that when the heart begins to beat again, the beats are small at first and gradually rise to normal. In some instances, however, they are more vigorous and quicker (Fig. 99).

3. Determine the Latent Period.—For this purpose a time-marker and an arrangement to indicate when the stimulus is thrown into the nerve are required.

(*a.*) Arrange the heart-lever as before, and adjust a time-marker to write exactly under the heart-lever.

(*b.*) Arrange an induction machine for an interrupted current, and keep Neef's hammer vibrating. Into the *secondary circuit* introduce an electro-magnet with a writing-lever attached to it; so adjust the electro-magnet that its writing-style writes exactly under the heart-lever, and arrange that when the writing-style on the electro-magnet is depressed—*e.g.*, by means of a weight—the secondary circuit is short-circuited, so that no stimulus is sent along the electrodes under the trunk of the vagus.

(*c.*) When all is ready lift the weight off the electro-magnet, whereby the secondary circuit is un-short-circuited, the electro-magnet lever rises up, records its movement on the cylinder, and at the same moment the induction shocks are sent through the vagus. Observe that the heart is not arrested immediately, but a certain time elapses—the *latent period*—usually about one beat of the heart (0·15 sec.) before the heart is arrested.

(*d.*) Short-circuit the secondary current again, and observe how the heart gradually resumes its usual rhythm, sinus venosus, auricles, and ventricle.

(*e.*) Repeat (*c.*) several times, noting that the heart after arrest goes on beating in spite of continued stimulation.

4. Gaskell's Method.—Instead of recording the movements of the heart by means of a lever resting on it, a very convenient method is that of Gaskell.

(*a.*) An ordinary writing-lever is placed above the frog-plate on which the frog rests, and supported in the horizontal position by a thin thread of india-rubber, as in Fig. 106. Expose the heart of a pithed frog, and leave it *in situ*. Tie a thread to the apex of the ventricle, clamp the aorta to fix the heart in position, and then attach the apex thread to the lever. Every time the heart contracts it pulls down the lever, and the latter is brought into position again, when the heart relaxes, by the piece of elastic.

5. Action of the Sympathetic on the Heart of the Frog.

(*a.*) Pith a frog or preferably a toad, cut away the lower jaw, and continue the slit from the angle of the mouth downwards for a short distance. Turn the parts well aside, and expose the vertebral column where it joins the skull. Remove the mucous membrane covering the roof of the mouth. The sympathetic is easily found before it joins the vagus emerging from the cranium (Fig. 100). Carefully isolate the sympathetic. It lies immediately under the levator anguli scapulæ, which must be carefully removed with fine forceps when the nerve comes into view, usually lying under an artery. Put a ligature round it as far away from the skull as practicable.

(*b.*) Expose the heart and attach its apex to a lever supported by an elastic thread as in Gaskell's method. Record several contractions, and then stimulate the sympathetic with interrupted shocks by means of fine electrodes. The heart beats quicker. If the heart is beating quickly, reduce the number of beats by cooling it.

(*c.*) If desired, the vagus may be isolated and stimulated, and the effects of the two nerves compared (although the vagus outside the skull is really the vago-sympathetic).

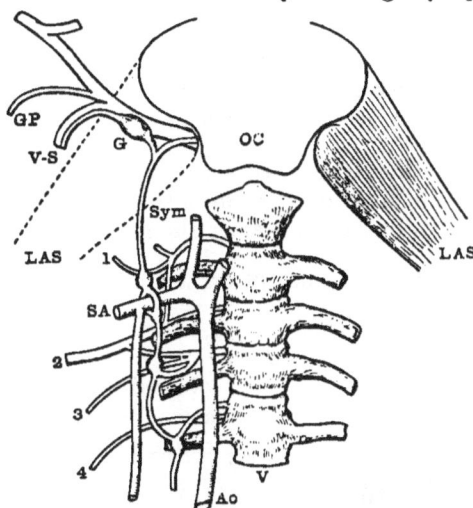

Fig. 100.—Scheme of the Frog's Sympathetic.—LAS, Levator anguli scapulæ ; Sym, sympathetic ; GP, glossopharyngeal ; V-S, vago-sympathetic ; G, ganglion of the vagus ; Ao, aorta ; SA, subclavian artery (*Gaskell*).

Stimulation of the intra-cranial vagus—*i.e.*, before it is joined by the sympathetic—is somewhat too difficult for the average student, and is therefore omitted here.

LESSON XLVIII.

ACTION OF DRUGS AND THE CONSTANT CURRENT ON THE HEART—DESTRUCTION OF THE CENTRAL NERVOUS SYSTEM.

1. Action of Drugs on the Heart—Muscarin and Atropin.— Either the excised heart, placed in a watch-glass, or the heart, *in situ*, may be used.

(*a.*) Pith a frog, expose its heart, and with a fine pipette apply a drop of serum or normal saline containing a trace of muscarin, which rapidly arrests the rhythmical action of the heart, the ventricle being relaxed—*i.e.*, in diastole—and distended with blood.

(*b.*) After a few minutes, with another pipette apply a few drops of a 0·2 per cent. solution of sulphate of atropia in normal saline, the heart gradually again begins to beat rhythmically.

(*c.*) Ligature and divide the frænum, raise the heart by the ligature, and faradise the crescent or inhibitory centre; the heart is no longer arrested, because the atropin has paralysed the intracardiac inhibitory mechanism.

(*d.*) In another frog, arrest the action of the heart with pilocarpin, and then apply atropin to antagonise it, observing that the heart beats again after the action of atropin.

2. Effect of a Constant Current on the Heart.

(*a.*) Expose a pithed frog's heart. Cut out the heart, dividing it below the auriculo-ventricular groove, thus obtaining an "apex" preparation which does not beat spontaneously.

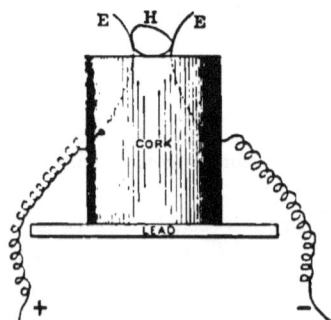

Fig. 101.—Support for Frog's Heart.—E, Electrodes; H, heart.

(*b.*) By means of sealing-wax, fix a cork to a lead base 5 cm. square, cover the upper end of the cork with sealing-wax, and thrust through it two wires to serve as electrodes, about 4 mm. apart (Fig. 101). Cover the whole with a beaker lined with moist blotting-paper. Place the heart-apex with its base against one electrode, and its apex against the other.

(*c.*) Arrange two Daniell's cells in circuit, connect them with a key, and to the latter attach the electrodes. Pass a continuous current in the direction of the apex. The heart resumes its rhythmical beating, and continues to do so as long as the constant current passes through the living preparation.

3. The Staircase.

(a.) To a glass slide used for microscopic purposes (3 × 1) fix with sealing-wax two copper wires in the long axis of the slide, and let their two free ends be about 3 millimetres apart. They will act as electrodes. Connect the other ends of the wires to a du Bois key introduced into the secondary circuit of an induction machine. Arrange the primary coil for single induction shocks, introducing a Morse key in the circuit.

(b.) Expose the heart of a frog, make an " apex preparation," and place it on the electrodes on the glass slide. Rest on the heart a heart-lever properly balanced and arranged to record its movements on a slow-moving drum (5 mm. per second). The preparation does not contract spontaneously, but responds to mechanical or electrical stimulation.

(c.) Stimulate the apex preparation with single break induction-shocks at intervals of about ten seconds. To do this un-short-circuit the secondary circuit, depress the Morse key, short-circuit the secondary circuit, and close the Morse key again. Repeat this several times every ten seconds, and note that the amplitude of the second contraction is greater than the first, that of the third than the second, the third than the fourth, and then the successive beats have the same amplitude (Fig. 102). Allow the heart-apex to rest for a few minutes, and repeat the stimulation. Always the same result is obtained. From the graduated rise of the first three or four beats after a period of rest, the phenomenon is known as the " staircase." The increment is not equal in each successive beat, but diminishes from the beginning to the end of the series.

Fig. 102.—Staircase character of Heart Beat.

(d.) If, while the apex is relaxing, it be stimulated by a

closing shock, it contracts again, so that the lever does not immediately come to the abscissa.

(*e.*) If the Morse key be rapidly tapped to interrupt the primary current, the contractions become more or less fused, and the lever remains above the abscissa writing a sinuous line.

4. Effect of Destruction of the Nervous System on the Heart and Vascular Tonus.

(*a.*) Destroy the brain of a frog, and expose its heart in the usual way, taking care to lose no blood; note how red and full the heart is with blood.

(*b.*) Suspend the frog, or leave it on its back, introduce a stout pin into the spinal canal, destroy the spinal cord, and leave the pin in the canal to prevent bleeding. Observe that the heart still continues to beat, but it is *pale* and *collapsed*, and apparently *empty*, it no longer fills with blood. The blood remains in the greatly dilated abdominal blood-vessels, and does not return to the arterial system, so that the heart remains without blood. If the belly be opened, the abdominal veins are seen to be filled with blood.

(*c.*) Amputate one limb, perhaps not more than one or two drops of blood will be shed, while in a frog with its spinal cord still intact, blood flows freely after amputation of a limb.

LESSON XLIX.

PERFUSION OF FLUIDS THROUGH THE HEART—PISTON RECORDER.

1. Perfusion of Fluids through the Heart.

The Fluid.—(*a.*) Take two volumes of normal saline, add one volume of defibrinated sheep's blood, mix, and filter. See that the blood is thoroughly shaken up with air before mixing it. This is the best fluid to use.

(b.) Rub up in a mortar 4 grammes of dried ox blood (this can be purchased) with 60 cc. of normal saline. Allow it to stand some time, add 40 cc. of water, and filter.

(c.) **Ringer's Fluid.**—Take 90 cc. of ·6 per cent. NaCl solution, saturate it with calcic phosphate, and add 10 cc. of a 1 per cent. solution of potassic chloride.

2. Preparation of the Heart.

(a.) Pith a frog, expose its heart, ligature and divide the frænum behind the ligature.

(b.) Take a two-wayed cannula (Fig. 103), attach india-rubber tubing to each tube, and fill the tubes and cannulæ with the fluid to be perfused. Pinch the india-rubber tubes with fine bull-dog forceps to prevent the escape of the fluid.

(c.) Tie a fine thread to the apex of the ventricle. To this thread a writing-lever is to be attached.

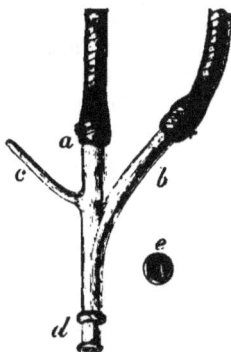

Fig. 103. — Cannula for Frog's Heart.

(d.) By means of the frænum ligature raise the heart, with a pair of fine scissors make a cut into the sinus, and through the opening introduce the double cannula passed through a cork, until its end is well within the ventricle. Tie it in with a ligature, the ligature constricting the auricles above the auriculo-ventricular groove, thus making what is known as a "heart-preparation." Cut out the heart with its cannula.

(e.) In a filter-stand arrange a glass funnel, with an india-rubber tube attached, at a convenient height, fill it with the perfusion fluid, clamp the tube. Attach this tube to one of the tubes—the inflow—connected with one stem of the cannula, taking care that no air-bubbles enter the tube. Adjust the height of the reservoir so that the fluid can flow freely through the heart, and pass out by the other tube of the cannula. Place a vessel to receive the outflow fluid. After a short time the heart will begin to beat.

(*f.*) Place the heart in a cylindrical glass tube, fixed on a stand, and arranged so that the cork in which the cannula is fixed fits into the mouth of the tube. A short test-tube does perfectly well. The lower end of the glass tube has a small aperture in it through which the thread (*c*) is passed, and attached to a writing-lever arranged on the same stand as the glass vessel. See that the lever is horizontal, and writes freely on a slow-moving recording drum. Every time the heart contracts it raises the lever, and during diastole the lever falls.

In this way it is possible to use various fluids for perfusion. The fluids may be placed in separate reservoirs, each communicating with the inlet tube, and capable of being shut off or opened by clamps, as required. Further, by poisoning the supply fluid with atropin, muscarin, spartein, or other drug, one can readily ascertain the effect of these drugs on the heart, or the antagonism of one drug to another.

Instead of a glass funnel as a reservoir for the fluid, one may use a Marriotte's flask (Fig. 104), the advantage being that the pressure of the fluid in the inflow tube is constant. Another simple arrangement is to have a bird's water-bottle, with a curved tube leading from it to the inflow tube of the cannula.

3. Piston-Recorder (of Schäfer).

The heart is tied to a two-way cannula as before, and is introduced into a horizontal tube with a dilatation on it. The tube of the recorder is filled with oil, and as the heart dilates it forces the oil along the tube and moves a light piston resting on it. When systole takes place, the oil recedes, and with it the piston. The piston records on a slow-moving drum placed horizontally.

LESSON L.

ENDOCARDIAL PRESSURE—APEX PREPARATION.

1. Endocardial Pressure in the Heart of a Frog.

(*a.*) Proceed as in the previous experiment (*a.*), (*b.*), (omit *c.*), (*d.*)

(b.) Arrange a frog's mercury manometer provided with a writing-style as in Fig. 104. Attach the inlet tube of the cannula to the Marriotte's flasks (a, b), and connect the outflow with the tube of the mercury manometer. It is well to have a T-tube between the heart and the manometer, but in the heart apparatus, as shown and used, the exit tube is preferable. See that no air-bubbles are present in the system. Every time the heart contracts the mercury is displaced and the writing-style is raised, and records its movements on a slow-moving drum.

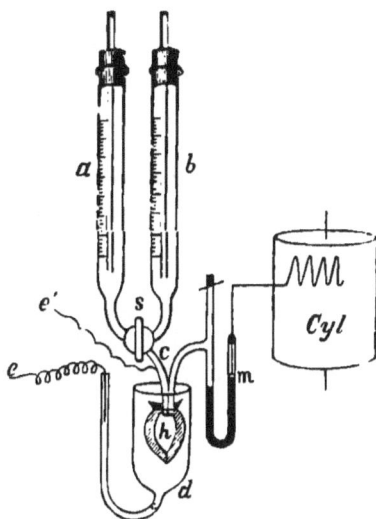

(c.) Take a tracing with the outflow tube and Marriotte's flask shut off, so that the whole effect of the contraction of the heart is exerted upon the mercury in the manometer. Take another tracing when the fluid is allowed to flow continuously through the heart. The second Marriotte's flask shown in the figure, is for the perfusion of fluid of a different nature, and by means of the stop-cock (s) one can pass either the one fluid or the other through the heart. The little cup (d) under the heart can be raised or lowered, and filled with the nutrient fluid, and in it the heart is bathed.

2. Apex Preparation.—In this preparation of the heart only the apex of the heart is used. As a rule, it does not beat spontaneously until sufficient pressure is applied to its inner surface by the fluid circulating through the heart.

(a.) Proceed as in Lesson XLIX., 2, (a.), (b.) (omit c.), (d.), with this difference, that in (d.) the cannula is placed deeper into the ventricle, and the ligature is tied round the ventricle below the auriculo-ventricular groove. Excise the heart and cannula, and attach it to the heart apparatus as in the previous experiment.

15

(*b.*) If the "heart-apex" preparation does not contract spontaneously, stimulate it by—*e.g.*, **single induction shocks**—either make or break. To this end adjust an induction machine, the wires from the secondary coil being attached, one to the cannula itself, while the other is placed in the fluid in the glass cup, into which the heart is lowered.

(*c.*) By introducing an electro-magnet with a recording lever into the primary circuit (Lesson XXXI.), and having a time-marker recording at the same time, one can determine the latent period of the apex preparation. It is about 0·15 sec.

(*d.*) If desired, the effect of a **constant current** may be studied in this way instead of by the method described in Lesson XLVIII., 2. The apex beats rhythmically under the influence of the constant current.

— -

LESSON LI.

TONOMETER—GASKELL'S CLAMP.

1. **Roy's Frog-Heart Apparatus or Tonometer.**—This apparatus registers the change of volume of the contracting heart. Fig. 105 shows a scheme of the apparatus. The apparatus consists of a small bell-jar, resting on a circular brass plate about 2 inches in diameter, and fixed to a stand adjustable on an upright. In the brass plate are two openings, the small one leads into an outlet tube (*e*), provided with a stop-cock. The other is in the centre of the plate, and leads into a short cylinder 1 cm. in length by 1 cm. in internal diameter. A groove runs round the outside of this cylinder near its lower edge, to permit of a membrane being tied on to it. In this cylinder works a light aluminum piston (*p*), slightly less in diameter than the cylinder. Around the lower aperture of the cylinder is tied a piece of flexible animal mem-

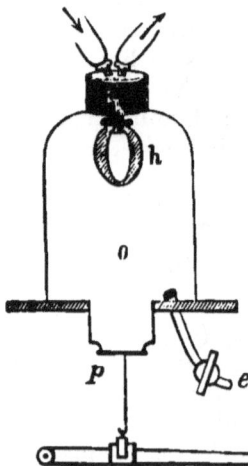

Fig. 105.—Roy's Tonometer.

brane, the ligature resting in the grooved collar. The free part of the membrane is tied to the piston, from the centre of whose under-surface (p) a needle passes down to be attached to a light writing-lever (l) fixed below the stage. The bell-jar is filled with oil (o), while in its upper opening is fitted a short glass stopper, perforated to allow the passage of a two-wayed heart-cannula with the heart attached (h). In using the instrument proceed as follows :—

(*a.*) Fix the bell-jar to the circular brass plate by the aid of a little stiff grease. Tie a piece of the delicate transparent membrane—such as is used by perfumers for covering the corks of bottles—in the form of a tube round the lower end of the grooved cylinder ; afterwards the lower end of the membrane is fixed to the piston, taking care that the needle attached to the piston hangs towards the recording lever. Drop in a little glycerin to moisten the membrane.

(*b.*) Fill the jar with olive oil, and have the recording apparatus ready, adjusted. Prepare the heart of a large frog [Lesson XLIX., (*a.*), (*b.*), (omit *c.*), (*d.*)], the cannula used being one fixed in the glass stopper of the bell-jar, and attach the inlet tube of the cannula to the reservoir of nutrient fluid, while the outlet tube is arranged so as to allow fluid which has passed through the heart to drop into a suitable vessel.

(*c.*) Introduce the cannula, with the heart attached, into the oil, and see that the stopper is securely fixed. Open the stop-cock (*e*), and allow some oil to flow out of O, thus rendering the pressure within sub-atmospheric; and as soon as the pressure has fallen sufficiently, and the little piston is gradually drawn up to the proper height, close the stop-cock. Attach the needle of the piston to the recording light lever, and take a tracing.

2. Gaskell's Clamp.

(*a.*) On a suitable support arrange two ordinary recording long, light levers of the same length, and with their writing-points exactly in the same vertical line, recording on a slow-moving drum, the levers being about 12 cm. apart. About midway between the two, place a Gaskell's clamp (Fig. 106, C), fixed in an adjustable arm attached to the same stand.

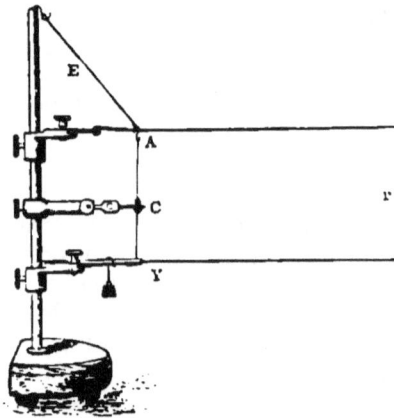

Fig. 106.—Gaskell's Clamp.

To support the upper lever, fix to it a fine thread of caoutchouc (E), and attach the latter to a slit or other arrangement on the top of the support. The clamp consists of two fine narrow strips of brass, like the points of a fine pair of forceps, which can be approximated by means of a screw.

(*b.*) Expose the heart of a pithed frog. Tie a fine silk thread to the apex of the ventricle, and another to the upper part of the auricles, and excise the heart. Tie the auricular thread to the upper lever and the ventricular one at a suitable distance to the lower lever.

(*c.*) Adjust the clamp (Fig. 106, C) so as to clamp the heart in the auriculo-ventricular groove, but at first take care not to tighten it too much, or merely just as much as will support the heart in position. After fixing the heart by means of the clamp, fix the two levers so that both are horizontal, and adjust the caoutchouc thread attached to the upper one, so that it just supports the upper lever, and when its elasticity is called into play by the contracting auricles pulling down the lever, it will, when the auricles relax, raise it to the horizontal position again.

(*d.*) Adjust a time-marker to write exactly under the writing-points of the two levers. Moisten the heart from time to time with serum or dilute blood.

(*e.*) After obtaining a tracing where the auricle and ventricle contract alternately, screw up the clamp slightly until the ratio of auricular to ventricular contraction alters—*i.e.*, until by compressing the auriculo-ventricular groove, the impulse from the auricles to the ventricle is "blocked" to a greater or less extent, when the auricles will contract more frequently than the ventricle.

<hr>

LESSON LII.

THE VALVES OF THE HEART—STETHO-SCOPE-CARDIOGRAPH—MEIOCARDIA AND AUXOCARDIA—REFLEX INHIBITION OF THE HEART.

1. **The Action of the Valves in the Dead Mammalian Heart.**— This is of value in order that the student may obtain a knowledge of the mechanical action of the valves. The heart and lungs of a sheep—with the pericardium still unopened—must be procured from the butcher.

(*a.*) Open the pericardium, observe its reflection round the blood-vessels at the base of the heart. Cut off the lungs moderately wide from the heart. Under a tap wash out any clots in the heart by a stream of water entering through both auricles. Prepare from a piece of glass tubing, 15 mm. in diameter, a short tube, 8 cm. in length, with a flange on one end of it, and another about 60 cm. long. Fix a ring to hold a large funnel on a retort stand.

(*b.*) Tie the short tube into the superior vena cava, the flanged end being inserted into the vessel. It must be tied in with well-waxed stout twine. In the pulmonary artery (P.A.)—separated from its connections with the aorta which lies behind it—tie the long tube, the flange securing it completely. Ligature the inferior vena cava, and the left azygos vein opening into the right auricle. Connect the short tube by means of india-rubber tubing with the reservoir or funnel in the retort stand. Keep the level of the water in the funnel below the upper surface of the P.A. tube. Fill the funnel with water; it distends the right auricle, passes into the right ventricle, and rises to the same height in the P.A. tube as the level of the fluid in the funnel. Compress the right ventricle with the hand, the fluid rises in the P.A. tube; and observe on relaxing the pressure that the fluid remains stationary in the P.A. tube, as it is supported by the closed semilunar valves. If the right ventricle be compressed rhythmically, the fluid will rise higher and higher until it is forced out at the top of the P.A. tube, and a vessel must be held to catch it. Observe that the column of fluid is supported by the semilunar valves, and above the position of the latter observe the three bulgings corresponding to the position of the sinuses of Valsalva.

(c.) Repeat (b.), if desired, on the left side, tying the long tube into the aorta, and the short tube into a pulmonary vein, ligaturing the others.

(d.) Cut away all the right auricle, hold the heart in the left hand, and pour in water from a jug into the tricuspid orifice. The water runs into the right ventricle, and floats up the tricuspid valves; notice how the three segments come into opposition, while the upper surfaces of the valves themselves are nearly horizontal.

(e.) With a pair of forceps tear out one of the segments of the semilunar valves of the pulmonary artery. Tie a short tube into the P.A., and to it attach an india-rubber tube communicating with a funnel supported on a retort stand. Pour water into the funnel, and observe that it flows into the right ventricle, floats up and securely closes the tricuspid valve. The semilunar valves have been rendered incompetent through the injury. Turn the heart any way you please, there is no escape of fluid through the tricuspid valve.

(f.) Take a funnel devoid of its stem, and with its lower orifice surrounded by a flange, and tie it into the aorta. Cut out the aorta and its semilunar valves, leaving a considerable amount of tissue round about it. Place the funnel with the excised aorta in a filter stand, and pour water into the funnel—much of it will escape through the coronary arteries, ligature these. The semilunar valves are quite competent—i.e., they allow no fluid to escape between their segments.

Hold a lighted candle under the valves, and observe through the water in the funnel how they come together, and close the orifice; observe also the triradiate lunules in apposition projecting vertically.

(*g.*) Slit open the pulmonary artery, and observe the form and arrangement of the semilunar valves.

(*h.*) Make a transverse section through both ventricles, and compare the shape of the two cavities, and the relative thickness of their respective walls.

(*i.*) Study two casts of the heart (after Ludwig and Hesse), (1) in diastole, and (2) in systole.

(*j.*) Ligature any large vessel attached to the heart, one feels the sensation of something giving way when the ligature is tightened. Cut away the ligature, open the blood-vessel, and observe the rupture of the coats produced by the ligature.

2. The Stethoscope—Heart Sounds.

(*a.*) Place the patient or fellow-student in a quiet room, and let him stand erect and expose his chest. Feel for the cardiac impulse, apply the small end of the stethoscope over this spot, and listen with the ear applied to the opposite end of the instrument. The left hand may be placed over the carotid or radial artery to feel the pulse in either of those arteries ; compare its time-relations with what is heard over the cardiac impulse.

(*b.*) Two sounds are heard—the *first* or *systolic* coincides with the impulse, and is followed by the *second* or *diastolic*. After this there is a pause, and the cycle again repeats itself. The first sound is longer and deeper than the second, which is of shorter duration and sharper.

(*c.*) Place the stethoscope over different parts of the præcordia, noting that the first sound is heard loudest at the apex beat, while the second is heard loudest at the second right costal cartilage at its junction with the sternum.

3. The Cardiograph.—Several forms of this instrument are in

use, including those of Marey (Fig. 107), Burdon-Sanderson, and the pansphygmograph of Brondgeest. Use any of them.

(*a.*) Place the patient on his back with his head supported on a pillow. Feel for the cardiac impulse between the fifth and sixth ribs on the left side, and about half an inch inside the mammary line.

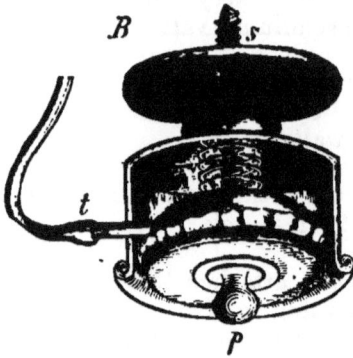

(*b.*) Arrange the cardiograph by connecting it (Fig. 107) with stout india-rubber tubing to a recording Marey's tambour adjusted to write on a drum (Fig. 76). It is well to have a valve or a T-tube capable of being opened and closed between the receiving and recording tambours, in order to allow air to escape if the pressure be too great.

Fig. 107.—Marey's Cardiograph.— *p*, Button placed over cardiac impulse; *s*, screw to regulate the projection of *p*; *t*, tube to other tambour.

(*c.*) Adjust the ivory knob of the cardiograph (*p*) over the cardiac impulse where it is felt most, and take a tracing. Fix, varnish and study the tracing.

4. Meiocardia and Auxocardia.

(*a.*) Bend a glass tube about 20 mm. in diameter into a semicircle, with a diameter of about 6 to 8 inches. Taper off one end in a glass flame to fit a nostril, and draw out the other end of the tube to about the same size. Round off the edges of the glass in a gas flame.

(*b.*) Fill the tube with tobacco smoke, place one end of it in one nostril, close the other nostril, cease to breathe, but *keep the glottis open.* Observe that the smoke is moved in the tube, passing out in a small puff during **auxocardia**—*i.e.*, when the heart is largest; while it is drawn further into the tube during **meiocardia**—*i.e.*, when the heart is smallest.

These movements, sometimes called the "cardio-pneumatic movements," are due to the variations of the size of the heart during its several phases of fulness, altering the volume of air in the lungs.

5. Reflex Inhibition of the Heart.

(*a.*) Place one hand over the chest of a rabbit and feel the beating of the heart. With the other hand suddenly close the nostrils of the rabbit, or bring a little ammonia near the nostrils, so as to cause the animal to close them. Almost at once the heart can be felt to cease beating for a time, but it goes on again.

6. Effect of Swallowing.

(*a.*) With your watch in front of you, count the number of your own pulse-beats per minute, and then slowly sip a glass of water, still keeping your finger on your pulse. Count the increase in the number of pulse-beats during the successive acts of swallowing. This is due to the inhibitory action of the vagus being set aside.

--- ---

LESSON LIII.

THE PULSE—SPHYGMOGRAPHS—SPHYG-MOSCOPE—PLETHYSMOGRAPH.

1. The Pulse.

(*a.*) Feel the pulse of a fellow-student, count the number of beats per minute; compare its characters with your own pulse, including its volume and compressibility. Observe how its characters and frequency are altered by exercise and by a prolonged and sustained deep inspiration.

2. The Sphygmograph.—Many forms of this instrument are in use. Study the forms of Marey and Dudgeon.

Marey's Sphygmograph (Fig. 108).

Mode of Application.—(*a.*) Cause the patient to seat himself beside a low table, and place his fore-arm on the double-inclined plane (Fig. 108), which, in the improved

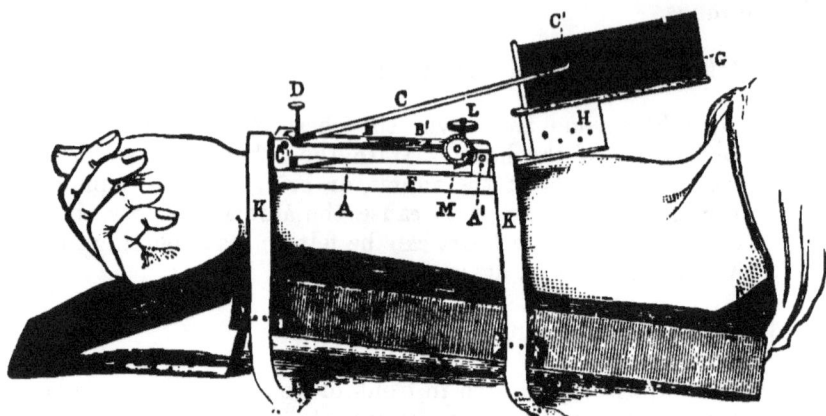

Fig. 108.—Marey's Sphygmograph applied to the Arm.

form of the instrument, is the lid of the box so made as to form this plane. The fingers are to be semiflexed, so that the back of the wrist, resting on the plane, makes an angle of about 30° with the dorsal surface of the hand.

(*b.*) Mark the position of the radial artery with ink or an aniline pencil. See that the clock (H) is wound up, and apply the ivory pad of the instrument exactly over the radial artery where it lies on the radius, and fix it to the arm by the non-elastic straps (K, K). The sphygmograph must be parallel to the radius, and the clock-work must be next the elbow. Cover the slide with enamelled paper, smoke it, fix it in position, and arrange the writing-style (C′) so as to write upon the smoked surface (G) with the least possible friction. Regulate the pressure upon the artery by means of the milled head (L)—*i.e.*, until the greatest amplitude of movement of the lever is obtained.

(*c.*) Set the clock-work going, and take a tracing. Fix it, scratch on the name, date, and pressure.

3. **Dudgeon's Sphygmograph** (Fig. 109).

(*a.*) Adjust the instrument on the radial artery by means

of an inelastic strap, carefully regulating the pressure—which can be graduated from 1 to 5 ounces—by means of

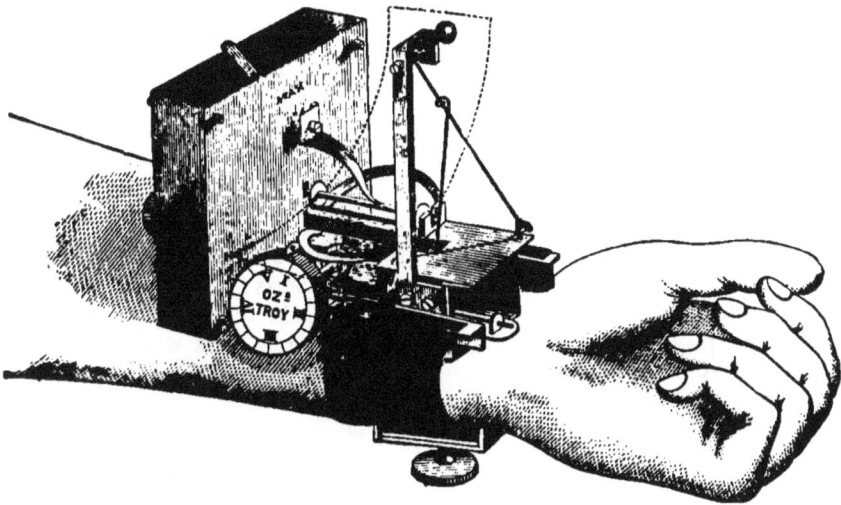

Fig. 109.—Dudgeon's Sphygmograph.

Fig. 109a.—Ludwig's Sphygmograph, made by Petzold, of Leipsig.

the milled head.   Smoke the band of paper, insert it between the rollers, and take a tracing.

4. Ludwig's Improved Sphygmograph.—Use this instrument (Fig. 109*a*).   It is not unlike a Dudgeon's Sphygmograph, but there is a frame adapted to the arm, and an arrangement for keeping the arm steady while the hand grasps a handle for the purpose.

5. Action of Amyl Nitrite.

(*a*.) With the sphygmograph adjusted, place *two drops— not more*—of amyl nitrite on a handkerchief, and inhale the vapour.   Within fifteen to thirty seconds or thereby it will affect the pulse, lowering the tension, the tracing presenting all the characters of a soft pulse-tracing, with a well-marked dicrotic wave.

6. The Gas Sphygmoscope (Fig. 110).

(*a*.) Connect the inlet tube of the instrument with the gas supply, light the gas-flame (*b*).   Apply the caoutchouc membrane (*a*) over the radial artery, and observe how the flame rises and falls with each pulse-beat.   Take a deep expiration, and observe the dicrotism in the gas-flame.

Fig. 110.—Gas Sphygmoscope, made by Rothe of Prague.

7. Plethysmograph.—Use the air-piston recorder of Ellis, and take a plethysmographic tracing of the variations of the volume of a finger.   The piston of the recorder must be lubricated with a volatile oil—*e.g.*, clove.

LESSON LIV.

THE PULSE-WAVE—RIGID AND ELASTIC TUBES—SCHEME OF THE CIRCULATION —RHEOMETER.

1. Velocity of the Pulse-Wave.

(*a.*) Take about 3 metres of india-rubber tubing about 6 mm. in diameter. To one end of the tube attach an ordinary elastic pump to imitate the heart, while the other end of the tube is left open, with a clamp lightly fixed on it. Arrange to pump water through the tube. Arrange two light levers on one stand, and place a part of the tube near the pump under the lower lever, and resting on a suitable support, while part of the tube near the outflow end is similarly arranged under the upper lever. Regulate the pressure of the lever upon the tube by means of lead weights.

(*b.*) Arrange on the same stand a Deprèz' chronograph to record the vibrations of an electro-tuning-fork, vibrating 30 or 50 D.V. per second. See that the writing-points of the two levers and the chronograph write upon the drum in the same vertical line.

(*c.*) Set the tuning-fork vibrating, allow the drum to move, compress the elastic pump interruptedly—to imitate the action of the heart—and propel water through the tube. The compression may be done by means of a lemon-squeezer, the extent of the excursion being regulated by a screw, and to secure regularity, arrange the number of pulsations to the beating of a metronome. On doing so, as one pumps in the water, the tube distends and raises the lever; in the interval between the beats, as the water flows out at the other end, the tube becomes smaller, and the levers fall. Feel the tube; with each contraction of the heart, a beat— the pulse-beat—can be felt.

(*d.*) Fix and study the tracing. The tracing due to the rise of the lever next the heart begins sooner, and is higher than the one from the lever near the outflow. Make two ordinates to intersect the three tracings, one where the lower pulse curve rises from the abscissa, and the other where the upper curve begins. Count the number of D.V. of the tuning-fork between these lines. Measure the length of the tube between the two levers, and from these data it is easy to calculate the velocity of the pulse-wave in feet per second.

2. Rigid and Elastic Tubes.—To the vertical stem of a glass T-tube or three-way tube 1 cm. in diameter, fix an elastic pump whose opposite end dips into a vessel of water. To the other slightly curved ends of the tube, fix a glass tube 90 cm. or thereby in length, and to the open end of the tube attach a small short piece of india-rubber tubing with a clamp over it. To the other limb attach an india-rubber tube of the same diameter and length as the glass tube, and fix a clamp over its outflow end. Arrange to pump water through the system. The pump may be compressed directly by the hand, or it may be placed between the two blades of a "lemon-squeezer," and the extent of the excursion of the latter regulated by a screw.

(*a.*) **Rigid Tube.**—Clamp off the elastic tube near the T-piece. Work the pump about 40 beats per minute, and force water into the glass tube. The water flows out *in jets* in an intermittent stream corresponding to each beat. Gradually clamp the outflow tube, and keep pumping, the water still flows out in an intermittent stream, and no amount of diminution of the outflow orifice will convert it into a continuous stream; as much water flows out as is forced in. All that happens is, that less flows out and, of course, less enters the tube. Instead of the clamp at the outflow, a tube drawn to a fine point may be inserted.

(*b.*) **Elastic Tube.**—Clamp off the glass tube near the T-piece, and unclamp the flexible one so as to have no resistance at its outflow end. Work the pump, the outflow takes place in jets corresponding to each beat of the pump. Pump as rapidly as possible, and the outflow stream will still be intermittent. While pumping, gradually clamp the tube at its outflow so as to introduce resistance there—to represent the resistance in the small arterioles—and when there is sufficient resistance at the outflow, the stream becomes a uniform and *continuous* one. Feel the tube; with each beat a pulse-beat is felt. The resistance at the periphery brings the elasticity of the tube into play between the beats, and thus converts the interrupted into a uniform flow. This apparatus serves also to demonstrate why there is no pulse in the capillaries, and under what circumstances a pulse is propagated into the capillaries and veins.

3. Scheme of the Circulation.—Use either Rutherford's scheme or the major schema. In the latter, the heart is represented by an ordinary elastic pump, the arteries by long elastic tubes dividing into four smaller tubes with clamps on them; two of the tubes leading into tubes filled with sponge to represent the capillaries. The capillaries lead into a tube with thinner walls representing the veins. The inflow tube into the heart, and the outflow tube at the vein are placed in a basin of water, and the whole system is filled with water.

(*a.*) Use two mercury manometers, and connect one with the arterial, and the other with the venous tube. Adjust a float on each, and cause the writing-points of the two floats to write exactly in the same vertical line on a revolving cylinder, the venous one a little below the arterial one.

(*b.*) Unclamp all the arteries, and work the pump, regulating the number of beats by means of a metronome beating 30 times per minute, and compress the heart to the same extent each time with a lemon-squeezer. Observe that both manometers oscillate nearly to the same extent with each beat. Take a tracing on a slow-moving drum.

(*c.*) Gradually clamp the arteries to offer resistance, and continue to work the pump, the pressure in the arterial manometer will rise more and more with each beat until it reaches a mean level with a slight oscillation with each beat. The pressure in the venous manometer rises much less, and the oscillations are very slight or absent.

(*d.*) While the mean arterial pressure is high, cease pumping; this will represent the arrest of the heart's action, brought about by stimulation of the peripheral end of the vagus—the arterial blood-pressure falls steadily.

(*e.*) Begin pumping again until the mean arterial pressure is restored, and then unclamp gradually the small arteries.

The steady fall of the blood-pressure represents the fall obtained when the central end of the depressor nerve is stimulated (the vagi being divided).

(*f.*) Two sphygmographs may be adjusted on the arterial tube, one near the heart, and the other near the capillaries, tracings being taken and compared.

(*g.*) The velocity of the pulse-wave may be estimated as in Lesson LIV., 1.

4. The Rheometer (Fig. 111) is used to measure the amount of blood flowing through a vessel in a given time. The nozzles of the instrument are inserted and tied into the artery of an animal, but as the student is not permitted to do this, use an india-rubber tube to represent the artery.

(*a.*) To represent the heart—or the weight of a column of fluid—arrange a Marriotte's flask or funnel on a stand, and to the outflow tube attach a narrow india-rubber tube, and clamp it after filling it with defibrinated blood. Suppose the tube to represent an exposed artery, clamp or apply and tighten two ligatures, about an inch apart, round the middle of the tube. Fill one bulb of the instrument with defibrinated blood and the other with clear almond oil, and close the top of the instrument with a glass plug.

(*b.*) Divide the part of the tube included between the two ligatures, and tie into either end the nozzles provided with the instrument. Call

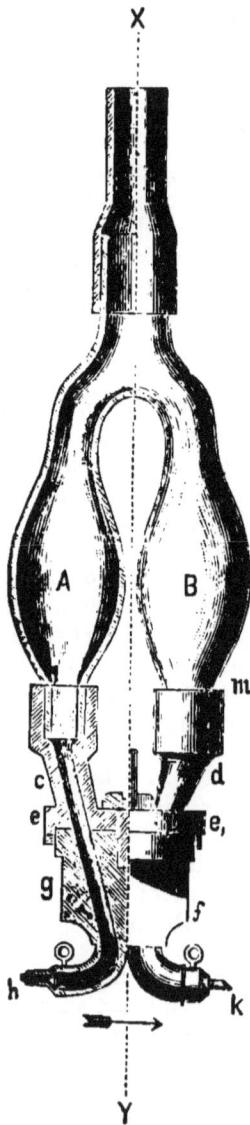

Fig. 111—Rheometer.

the one next the reservoir or heart (h), and the other one (k). Fix the instrument into the nozzles, the bulb A being filled with oil, and in connection with h, B with defibrinated blood, and connected with k. The instrument is fixed in position by a support provided with it, while a handle which fits into two tube-sockets on the upper surface of the disc (e, e_1) is used to rotate the one disc on the other.

(*c.*) All being now ready, take the clamp off the reservoir of blood and the clamps or ligatures off the artery. The defibrinated blood flows into the bulb A, displaces the oil in it towards B, the defibrinated blood of B being forced out into the artery and caught in a suitable vessel. Of course, in the animal this blood simply passes into the upper end of the artery. As soon as the bulb A is filled with blood, which is indicated by a mark on the glass, the disc is suddenly rotated, whereby B communicates with h, and A with k. The blood now flows into B, displacing the oil in it into A, and as soon as this takes place, the disc is again rotated. This process is repeated several times. Count the number. The bulbs have the same capacity and are exactly calibrated, so that to measure the amount of blood flowing through the blood-vessel we require to know the time occupied.

The time is most conveniently measured by connecting the rheometer with an electro-magnet registering on a drum each rotation of the disc, and under this a time-marker records seconds, or a tuning-fork may be used.

LESSON LV.

BLOOD-PRESSURE AND KYMOGRAPH— CAPILLARY BLOOD-PRESSURE— LYMPH-HEARTS.

1. Estimation of the Blood-Pressure by Ludwig's Kymograph.— As students are not permitted to perform experiments upon live animals, the most they can do in this experiment is to arrange the necessary apparatus as for an experiment, and to make the necessary dissection on a dead animal.

16

A. (*a*.) **Arrange the recording apparatus** for a continuous
tracing. The clock-work is wound up, and the drum is so
adjusted, that, when it moves, it unwinds the continuous
white paper from a brass bobbin placed near it. Arrange
a time-marker connected with a clock, provided with an
electric interrupter, to mark seconds at the lower part of the
paper. It is usual to use a pen-writer charged with a
solution of aniline to which a little glycerin is added to
make it flow freely.

(*b*.) Partially fill the manometer with dry clean mercury,
and in the open limb of the manometer place the float,
provided with a pen or sable brush moistened with aniline
ink containing a little glycerin. See that the float rests
on the convex surface of the mercury.

(*c*.) The closed or proximal side of the manometer at
its upper part is like a T-tube, the stem of which is
connected by thick india-rubber tubing to a piece of
flexible lead tubing, on the free end of the latter is tied a
glass cannula of considerable size, and over the india-rubber
tubing connecting the cannula with the lead tube is placed
a clamp. The proximal end of the manometer is filled
by means of a pipette with a saturated solution of sodic
carbonate as high as the stem of the T-piece. To it is
attached a long india-rubber tube, which is connected with
a pressure-bottle filled with a saturated solution of sodic
carbonate, and kept in position by a cord passing over a
pulley fixed in the roof. A clamp compresses the india-
rubber tube just above the manometer. Open this clamp
and also the one at the end of the lead pipe. The alkaline
solution fills the whole system, and after it does so, and no
air-bubbles are present, close the clamp at the end of the
lead tube, and then the one on the pressure-bottle tube.
It is well to have an inch or more of positive pressure in
the manometer. See that the writing-style writes smoothly
on the paper, and that it is kept in contact with the latter
by a silk thread with a shot attached to its lower end.

B. **To Insert the Cannula.**—(*a*.) **Arrange the necessary
instruments** in order on a tray. Scissors, scalpels, forceps
(coarse and fine), seeker, well waxed ligatures, small aneurism needle, bull-dog forceps, cannulæ, sponges.

(b.) Make the necessary dissection on a dead rabbit. Fix the rabbit in a Czermak's holder as would be done if the animal were alive. Clip away with a pair of scissors the hair over the neck, and with a moist sponge moisten the skin to prevent any loose hair from flying about. Pinch up the skin on one side of the trachea, between the left thumb and forefinger, and divide it with a sharp scalpel. This exposes the fascia, which is then torn through with forceps, draw the sterno-mastoid aside, and gently separate the muscles with a "seeker" until the carotid, accompanied by the vagus, depressor, and sympathetic nerves is seen.

(c.) Open the sheath, and with the seeker carefully isolate about an inch of the carotid. Pass a ligature under the artery by means of a fine aneurism needle, withdraw the needle, and ligature the artery. About an inch on the cardiac side of the latter, clamp the artery with bull-dog forceps. Raising the artery slightly by the ligature, with a fine-pointed pair of scissors make an oblique V-shaped slit in the artery, and into it introduce a suitable glass cannula with a short piece of india-rubber tubing tied on to it. Place another ligature round the artery, and tie it round the artery and over the shoulder of the cannula. The point of the cannula is of course directed towards the heart. Fill the cannula with the soda solution, and into the cannula slip the glass nozzle at the end of the lead pipe, tying it in securely. Unscrew the clamp at the end of the elastic tubing. Set the clock-work going; if one were operating on a living animal, the next thing to do would be to remove the clamp or forceps between the artery cannula, and the heart. At once the swimmer would begin to move and record its oscillations on the paper moving in front of it.

(d.) Before joining the lead tube to the cannula, isolate the vagus, the largest of the three nerves, put a ligature round it, and divide it above the ligature. Isolate also the depressor nerve, put a ligature round it low down in the neck, and divide it between the ligature and the heart. The latter is easily distinguished from the sympathetic, as it is the smallest of the three nerves accompanying the carotid. In the dead rabbit the depressor may be traced up to its origin by two branches, one from the vagus, and the other from the superior laryngeal. Moreover, if the sympathetic be traced upwards, a ganglion will be found on it. This is merely to be regarded as an exercise for practice.

(e.) In every case a base-line or line of no pressure must be recorded on the continuous paper. This indicates the abscissa, or when the mercury is at the same height in the two limbs of the manometer.

(f.) Measure a blood-pressure tracing provided for you. Lay the tracing on a table. Take a right-angled triangle made of glass or wood, and place one of the sides bounding its right angle upon the abscissa, the other side at right angles to this has engraved on it a millimetre scale. Read off the height in millimetres from the base-line to the lowest point in the curve, and also to its highest point, take the mean of the two, and multiply by two, this will give the *mean arterial pressure*. Instead of measuring only two ordinates, measure several, and take the mean of the number of measurements. In all cases the result has to be multiplied by two.

(g.) Study blood-pressure tracings obtained by stimulation of

(i.) The peripheral end of the vagus.

(ii.) The central end of the depressor.

(iii.) The central end of a sensory nerve.

(h.) In every kymograph tracing, notice the smaller undulations due, each one, to a single beat of the heart, and the larger ones due to the respiratory movements. In a blood-pressure tracing taken from a dog with the vagi not divided, observe that the size of the heart-beats on the descent of the respiratory wave is greater, while the number of beats is less than on the ascent.

2. To Make a Cannula.—Heat in the flame of a blow-pipe a piece of hard glass tubing about 5 mm. in diameter. When it is soft, take it out of the flame, draw it out gently for about 3 cm. Allow it to cool ; make the gas jet smaller, heat the thin drawn-out part of the tube, and draw it out very slightly. This makes a shoulder. With a triangular file just scratch the narrow part obliquely beyond the second constricted part, and break it off. A cannula with a shoulder and an oblique narrow orifice is thus obtained. Round off the oblique edges either by a file, rubbing

them on a whetstone, or heating slightly in a gas-flame.   Tie a
piece of india-rubber on the other end, and the cannula is
complete.

3. Blood-Pressure in the Capillaries.

(a.) Make the following apparatus (Fig. 112), consisting
of a slip of glass, 2 cm. long, 3 to 4 mm. broad, and 1 mm.
thick, and on its under surface fix with cement a glass plate
(a), with a surface of 5 mm. square.   Two threads supporting
a paper scale-pan are attached to the glass plate.   Place the
glass plate (a) over the skin on the dorsal surface of the
finger, just at the root of the nail.   Add weights to the
scale-pan until the skin becomes pale.   Note the weight
necessary to bring this about, but observe that the skin does
not become pale all at once.

(b.) Test how altering the position of the hand affects the
pressure in the capillaries.

Fig. 112.—Apparatus
used by v. Kries for
estimating the capil-
lary blood-pressure.

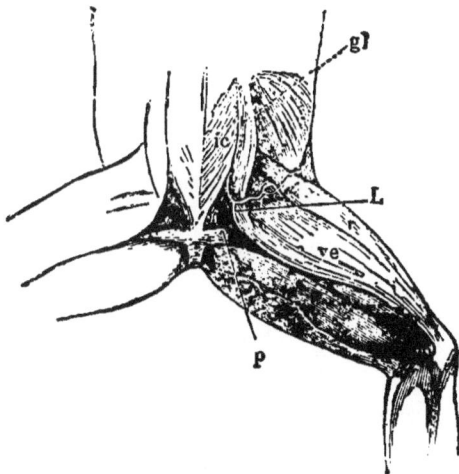

Fig. 113.—Posterior Pair of Lymph-Hearts
(L) of the Frog.

4. Examine with a microscope the circulation in the web of a
frog's foot, or in the mesentery of a frog with its brain destroyed,
and afterwards slightly curarised.

5. The Lymph-Hearts.

(*a.*) Destroy the brain of a frog, place it on its belly, and watch the beating of the posterior pair of lymph-hearts, which are situated one on each side of the urostyle (Fig. 113).

(*b.*) Remove the skin covering them, taking care not to cut too far outwards, else a cutaneous vein will be injured and bleed freely. Count the number of beats per minute, noting that the rhythm is not synchronous with the blood-heart, whose movements can usually be distinguished without opening the chest.

(*c.*) Destroy the posterior part of the spinal cord with a seeker or wire, and observe that the rhythmical automatic movements of the lymph-hearts cease.

PHYSIOLOGY OF RESPIRATION.

LESSON LVI.

MOVEMENTS OF THE CHEST WALL— ELASTICITY OF LUNGS— HYDROSTATIC TEST.

1. Movements of the Chest Walls—Stethograph.

A. Rabbit.—(*a.*) Arrange a drum and time-marker. Fix a rabbit conveniently, and with tapes tie on its chest Marey's double-tambour, connecting the latter with a recording tambour adjusted to write on the drum. Introduce between the receiving and recording tambour either the valve usually supplied with Marey's apparatus, or a T-tube with a screw clamp, whereby the pressure within the system of tubes can be regulated. Take a tracing. If one of the receiving tambours be placed over the cardiac impulse, the tracing will show also the number of beats of the heart.

B. Man—Stethograph (Marey).—(*a.*) Cause a person to expose his chest. Raise the screw (*g*) of the stethograph, and fix the plate (*f*) of the instrument on the chest, with tapes attached to (*c*) and (*d*). Depress (*g*), connect the tube (*a*) with a recording tambour, with the same precautions as in 1, **A**, and take a tracing. Examine the tracing, noting the relation between inspiration and expiration.

2. Stethometer of Burdon-Sanderson.

(*a.*) Prepare a drum and time-marker, as in the previous experiments. Cause the person to expose his chest, and seat himself conveniently. The instrument is suspended by

a broad band placed round the neck, the horizontal bar being
behind the body.

(b.) The most important diameters of the chest to measure
are—" Those connecting the eighth rib in the axillary line
with the same rib on the opposite side, the manubrium
sterni with the third dorsal spine, the lower end of the
sternum with the eighth dorsal spine, and the ensiform

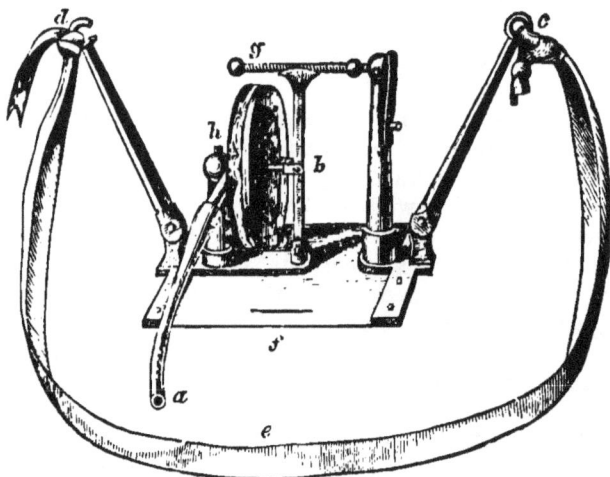

Fig. 114.—Marey's Stethograph.

cartilage with the tenth dorsal spine." Measure only the
first. Adjust the knob of the tambour on one side against
the eighth rib, as above, while the movable bar with its
knob is placed against the opposite corresponding rib. Connect the tambour with the recording tambour, introducing
a T-piece, the stem of which is provided with an india-
rubber bag and screw clamp to regulate the pressure within
the air-system.

3. Intra-Thoracic Pressure.—For practice this can be done on
a dead rabbit.

(a.) Fix the dead rabbit in Czermak's rabbit-holder.
Expose the trachea, tie into it a knee-shaped glass cannula.
Make a small water-manometer or bent U-tube, with a

millimetre scale attached, fill it about half full with coloured water, and to the proximal limb attach an india-rubber tube with a **T**-piece and screw clamp, as in the other experiments. Connect the tracheal cannula with the manometer tube, tighten the screw clamp, and see that the water stands at the same level in both limbs of the manometer.

(*b.*) Open both pleuræ without injuring the lungs. The lungs collapse, and the water is depressed in the proximal side of the manometer, and rises in the open limb.

4. Elasticity of the Lungs.

(*a.*) Remove the whole of the front of the chest in the rabbit already used. Observe the collapsed lungs. To the tracheal cannula attach an india-rubber bag such as is used with a spray-producer, and inflate the lungs. Cease to pump air into the lungs, and observe how they collapse.

5. Hydrostatic Test.

(*a.*) Cut out the lungs and the heart. Place them in a vessel of water. The whole will float, as the lungs contain so much air. Cut off a small piece of one lung, throw it into water, it floats. This is the **hydrostatic test.** Compare a piece of pneumonic lung, the latter sinks.

6. **Apnœa.**—Count the number of your own respirations per minute. Take a series of rapid inspirations. Note that several seconds elapse before the next inspiration. This is the period of apnœa.

LESSON LVII.

VITAL CAPACITY—EXPIRED AIR— LARYNGOSCOPE—VOWELS.

1. **Vital Capacity.** — Estimate this on Hutchinson's spiro- meter—*i.e.*, take the deepest possible inspiration, and then make the deepest possible expiration, expiring into the mouthpiece of the spirometer.

2. Changes in Expired Air.

(*a*.) **Black's Experiment.**—Place equal quantities of lime-water in two vessels (A and B). Take a deep breath, close the nostrils, and expire through a bent glass tube into A. The lime-water soon becomes milky, owing to the large amount of carbonic acid expired, combining with the lime to form carbonate of lime. With the elastic pump of a spray-producer, pump the air of the room through B. B remains clear and does not become turbid. Therefore, the carbonic acid must have been added to the inspired air in the respiratory organs.

(*b*.) **Müller's Valves.**—Arrange two flasks (A and B) and tubes as in Fig. 115, with some lime-water in both. Close

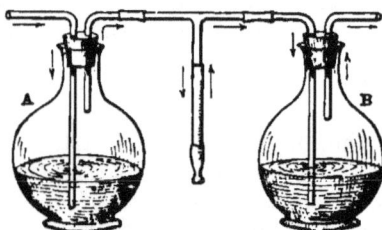

Fig. 115.—Müller's Valves.

the nostrils, apply the mouth to the tube, and inspire. The air passes in through A, and is freed of any CO_2 it may contain. Expire, and the air goes out through B, in which the lime-water becomes turbid.

(*c*.) **Heywood's Experiment.**—Place about two litres of water in a basin, and in it put erect a bell-jar without a bottom. Ascertain that a lighted taper burns in the jar. Renew the air, place in the neck of the latter a glass tube with a piece of india-rubber tubing attached. Close the nostrils, apply the mouth to the tube, and inspire. The water rises in the bell-jar. Then expire, the water sinks, and the air which was originally present above the water has been taken into and expelled again from the respiratory passages. Remove the cork, and place a lighted taper in the expired air. The taper is extinguished (Fig. 116).

3. Study the construction and mode of using a **gas-pump** for the analysis of blood-gases.

4. The laryngoscope is used to investigate the condition of the pharynx, larynx, and trachea. Various forms are in use, but they all consist of—(1) One or more small usually circular plane mirrors fixed to a metallic rod at an angle of 120°, the metallic rod fits into a suitable handle, and is fixed by means of a screw.

Fig. 116.—Heywood's Experiment.

(2) A large concave mirror of about 20 cm. focus perforated with a hole in the centre, and secured to the operator's forehead by means of a circular band passing round the head. The mirror itself is fixed in a ball and socket joint so that it can be moved freely in every direction.

A. Practise first of all on a **model** of the head and larynx provided for the purpose.

B. On a Living Person.—(*a.*) Place the patient upright in a chair. A good source of artificial light—*e.g.*, a suitable Argand lamp—is placed near the side of the patient's head, a little above the level of his mouth. The new incandescent lamp gives a brilliant, clear, and steady light. M. Mackenzie's rack-movement lamp is a most convenient form. The observer seats himself opposite and close to the patient; places the large mirror on his forehead, and either looks through the central hole in it with one eye, or raises it so that he can just see under its lower edge.

(*b.*) Seated in front of the patient, he directs a beam of light until the lips of the patient are brightly illuminated. The patient is then directed to incline his head slightly backwards, to open his mouth wide, and protrude his tongue. Place a clean handkerchief over the tongue, and give the patient the handkerchief to hold, which secures that the tongue is kept protruded and well forward. Move the large mirror until the uvula and back of the throat are brightly illuminated, the operator moving his head slightly to and from the patient until the greatest brightness is obtained.

(*c.*) Take the small laryngeal mirror in the right hand, and warm it gently over the lamp to prevent the condensation of moisture on its surface. Test its temperature on the skin of the cheek or the back of the hand. Holding the handle of the mirror as one does a pen, rapidly carry it horizontally backwards, avoiding contact with any structures in the mouth, until its back rests against the base of the uvula. At the same time, direct the beam of light upon the laryngeal mirror, when an inverted image of the larynx will be seen more or less perfectly.

(*d.*) By moving the laryngeal mirror, not, however, pressing too much on the uvula, or continuing the observation for too long a time, one may explore the whole of the larynx. Perhaps only the posterior part of the *dorsum of the tongue* is seen at first; if so, slightly depress the handle of the mirror, when the curved fold of the slightly yellowish *epiglottis* and its cushion, with the *glosso-epiglottidean folds*, come into view. In the middle line are the *true vocal cords*, which are pearly white and shining, and best seen when a high note is uttered, and between them the chink of the *glottis*. Above these are the *false vocal cords*, which are red or pink, the *ary-epiglottidean folds*, with on each side the *cartilages of Wrisberg* furthest out, the *cartilages of Santorini* internal to this, and the *arytenoid cartilages* near the middle line.

(e.) Make the patient sing a deep or high note, or inspire feebly or deeply, and observe the change in the shape of the glottis. On uttering a deep note, the rings of the trachea may be seen. *N.B.*—Remember that what is seen by the observer in the laryngeal mirror on his right or left corre-

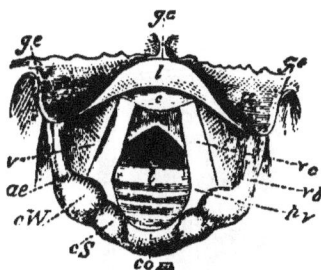

Fig. 117.—View of the Larynx during a Deep Inspiration.—*g.e*, Glosso-epiglottidean fold; *l.c*, lip and cushion of epiglottis; *a.e*, ary-epiglottic fold; C.W, C.S, cartilages of Wrisberg and Santorini; *v.c*, vocal cord; *v.b*, ventricular band; *p.v*, processus vocalis; *c.r*, cricoid cartilage; *t*, rings of trachea.

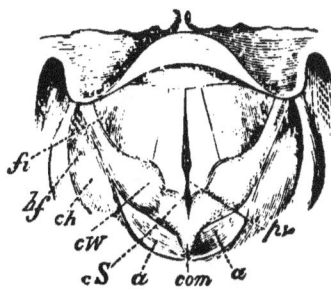

Fig. 118.—Larynx during Vocalisation.—*f.i*, fossa innominata; *h.f*, hyoid fossa; *com*, arytenoid commissure.

sponds to the *patient's* left and right. The lower part of the mirror gives an image of the more posterior structures, while the anterior structures are reflected in its upper part.

5. Auto-Laryngoscopy.—The student should learn to use the laryngoscope on himself. The student sits in a chair, fixes the large reflecting mirror in a suitable holder about eighteen inches in front of, and on a level with his mouth. Behind and to one side of this an ordinary plane mirror is placed vertically. On one side of his head he places the source of light. The light is reflected on to the uvula by the reflecting mirror, and on introducing the small laryngeal mirror, by a little adjustment, one sees the image of the larynx in the plane mirror. Or one may use in a similar way—the apparatus of Foulis. In Dr. George Johnson's method, the ordinary reflector is strapped on to the forehead, and the observer places himself in front of a toilet mirror. In a line with and slightly behind the mirror, and on one side of the observer, place a lamp. By means of the reflector, the image of the fauces seen in the mirror is illuminated. Introduce the laryngeal mirror, when the image of the larynx is seen in the toilet mirror.

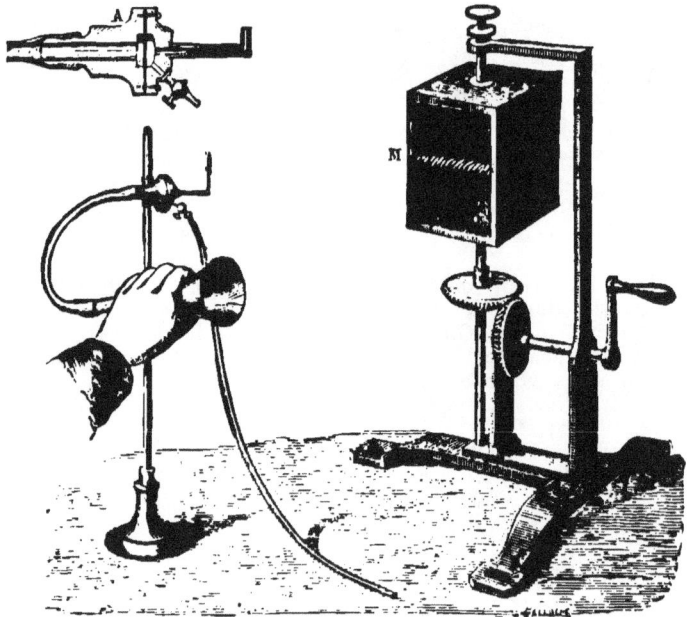

Fig. 119.—Koenig's Manometric Flame Apparatus.

6. Analysis of Vowel Sounds.

(a.) Use Koenig's apparatus, as shown in Fig. 119. Connect the tube of the capsule with the gas supply, light the gas jet, and sing the vowels A, E, I, O, U in front of the open trumpet-shaped tube shown in the figure. With the other hand rotate the mirror (M), and observe the serrated reflection of the flame in the mirror, noticing how the image in the mirror varies with each vowel sounded.

PHYSIOLOGY OF THE CENTRAL NERVOUS SYSTEM.

REFLEX ACTION—ACTION OF POISONS— KNEE-JERK.

1. **Reflex Action.**—Destroy the brain of a frog, which should be done without loss of blood. Place under a bell-jar a normal frog for comparison. Observe that immediately the frog is pithed, on pinching one of its toes, very probably the leg will not be drawn up. Allow the frog to rest for half an hour or more, and observe—

(*a.*) Its *attitude;* the head of the.pithed frog lies on the plate on which it is placed, while in the intact frog, the head is erect, the body and head forming an acute angle with the surface on which the frog rests.

(*b.*) Its eyes are closed, while those of the intact frog are open. The fore limbs are either flexed and drawn under the chest, or spread out, so that the body is no longer supported on the nearly vertical fore limbs as in the intact frog, but lies flat upon the surface of support. The legs are pulled up towards the body.

(*c.*) The absence of respiratory movements in the nostrils and throat. Observe that it makes no spontaneous movements, if left entirely to itself.

(*d.*) Turn it on its back, it lies in any position it is placed. Do this with a normal frog, the latter regains its equilibrium at once. Pull out one of the legs, it will be

drawn up again towards the body. Pinch the flank with a
pair of forceps; the leg of the same side is rapidly extended,
then drawn up towards the spot stimulated.   Pinch sharply
the skin round the anus with forceps. Immediately, both
legs are pushed out and pulled up towards the body, as if to
dislodge the offending body.

2. Bend a long (6 cm.) straight pin into the form of a hook,
and push it through the tips of both jaws, and by means of the
hook hang up the frog on a suitable support. It hangs vertically
with the legs pendant. At first the legs may make a few movements, but they soon cease to do so, and hang quite motionless.

(*a.*) Pinch the tip of any toe of the right leg, the right
leg is flexed and drawn up. If a toe of the left leg be
pinched, either with the nail or forceps, the left leg is
drawn up.

3. The Latent Period (Türck's Method).

(*a.*) Prepare dilutions of sulphuric acid containing 1, 2,
3, and 4 cc. per litre—*i.e.*, 0·1, 0·2, 0·3, and 0·4 per cent. of
sulphuric acid, and place some of each in four shallow
glasses. Arrange also a large beaker of water to wash the
frog. Adjust a metronome to beat one hundred times per
minute. Cause it to beat.

(*b.*) Hold the frog in the left hand by means of the hook,
and in the right take a glass rod to hold one leg aside.
Dip the other leg up to the ankle into the 0·1 per cent. acid,
and on doing so count the number of beats before it is
withdrawn from the acid. After the leg is withdrawn,
wash the leg in water to remove the acid. Note the time
in hundredths of a minute—*i.e.*, the **latent period**. Allow
the frog to rest at least five minutes, and repeat the experiment. Take the mean of the two observations—or if you
prefer it of three or more observations—and this will give
the "latent period."

(*c.*) Repeat with suitable intervals of repose the same
experiment with acid of 0·2, 0·3, and 0·4 per cent., noting
that as the strength of the acid increases, the latent period
becomes shorter, but not in the ratio in which the acid is
stronger.

4. Chemical Stimulation.

(*a.*) In a small glass place some strong acetic acid and a few pieces of filter paper 3 mm. square. Either when the frog is lying on its back, or while it is suspended, apply with a pair of forceps one of the pieces of paper moistened with acid—the surplus removed—to the skin on the inner side of the thigh. At once the leg on that side is violently drawn up, perhaps both legs are drawn up, and the foot of the leg first drawn up is swept over the spot stimulated, as if to remove the piece of paper—*i.e.*, purposive, well-co-ordinated movements are executed. At once dip the frog in water to remove the acid, allow it to rest for some time.

(*b.*) After an interval of five minutes repeat the experiment, but hold the leg to which the acid is applied. In all probability the other leg will be moved, and the opposite foot will be used to remove the irritating acid paper. Wash the frog and allow it to rest.

(*c.*) Test further by applying papers to the flank, the skin over the gastrocnemius, &c., and in all cases characteristic, but different reflex movements will be elicited, if sufficient interval for recovery (5 mins. at least) be allowed between the successive experiments.

(*d.*) With a needle destroy the spinal cord, all reflex action is then abolished, although the nerves and muscles retain their excitability, and the heart continues to beat. Expose the heart and observe it beating. Test the excitability of the nerves and muscles in the usual way with an interrupted current.

5. Action of Strychnia.

(*a.*) Using a frog with its brain destroyed, inject with a fine glass pipette or a hypodermic syringe into the dorsal lymph-sac a drop of dilute solution of sulphate of strychnia (0·5 per cent.)

(*b.*) Observe, that as soon as the poison is absorbed—*i.e.*, within a few minutes—cutaneous stimulation of any part of the body, even tapping the table excites violent tetanic

17

spasms—or general tetanus—of the whole body. During the paroxysm of convulsions, the limbs are stretched out, extended, hard, and rigid, while the trunk is similarly affected. The extensor muscles are more affected than the flexors. The tetanic paroxysm passes off, to be soon followed by another on the slightest stimulation.

(c.) Destroy the spinal cord with a seeker or long pin. At once the spasms cease. Strychnia, therefore, acts on the cord directly, and not on the muscles and nerves.

6. Action of Potassic Chloride or Bromide.

(a.) Prepare a reflex frog as in Lesson LVIII., 1. Test the latent period with dilute sulphuric acid, 0·2 per cent., until constant results are obtained. Inject 2 minims of a 1 per cent. solution of KCl or KBr, and after ten minutes test again the latent period. Within a short period the latent period will be greatly prolonged.

7. Knee-Jerk.

(a.) Sit on a chair and cross the right leg over the left one. With the tips of the fingers, or a percussion hammer, strike the right ligamentum patellæ. The right leg will be raised and thrown forward with a jerk owing to the contraction of the quadriceps muscle. An appreciable time elapses between the striking of the tendon and the jerk. The knee-jerk is almost invariably absent in cases of locomotor ataxia, while it is greatly exaggerated in some other nervous affections, so that its presence or absence is a most important clinical symptom.

LESSON LIX.

NERVE ROOTS—REACTION TIME.

1. Functions of the Roots of the Spinal Nerves.—To expose the roots destroy the brain of a frog, lay it on its belly, and make a median incision in the skin of the back, from the neck to the upper end of the urostyle. Turn back the flaps of skin, and carry the incision down to the spines of the vertebræ. With a scraper or blunt knife remove the muscles along each side of the vertebral column, so as to lay bare the arches of the vertebræ. With a blunt-pointed pair of scissors, or two saw blades parallel to each other and fitted at a suitable distance into a handle, as devised by Ludwig, cut through the arches of the eighth or last vertebra, taking care not to injure the nerves within the spinal canal. Remove successively from below upwards the seventh, sixth, and fifth vertebral arches, when the tenth, ninth, and eighth spinal nerve roots will come into view. The posterior roots are larger, come first into view, and cover the anterior. The roots may be separated by a seeker. Select the largest posterior root —the ninth—and with an aneurism needle carefully place a fine silk thread (say a red one) under it.

(a.) Tighten the ligature near the cord, and observe movement in some part of the body. Divide the nerve between the cord and the ligature, and observe further movements on division.

(b.) With the thread gently lift up the *peripheral* or distal end of the nerve root, place it on well-protected electrodes, and stimulate it with an interrupted current. No movement is observed in the muscles of the limb.

(c.) Select the posterior root of the eighth nerve, ligature it at some distance from the cord, and divide it on the distal side of the ligature. There is neither contraction of the muscles of the leg nor movement of the body. Place the *central* stump—*i.e.*, the part still connected with the cord on the electrodes, and stimulate it, when movements will take place in several parts of the body.

(d.) Divide the posterior roots of the seventh and tenth nerves. Observe that the whole limb on that side has become insensible. Turn aside the roots of the divided nerves, and expose the anterior roots, which are very thin and slender. Repeat the preceding experiments on the anterior root of the ninth nerve—*i.e.*, place a ligature around it, tighten the ligature, and divide the nerve between the cord and the ligature. Stimulate the *distal* end with an interrupted current; this causes contraction of the muscles of the limb supplied by this root.

(e.) Repeat the experiment on the eighth nerve root. Stimulate the *central* end, no effect is produced.

From the effects of section and stimulation of the nerve roots, one concludes that the anterior are motor, and the posterior are sensory.

2. **Reaction Time**—*i.e.*, the interval between the application of a stimulus to a sense-organ and the moment the stimulus is responded to by the individual. The **Neuramoebometer** (*Exner*), or **Psychodometer** (*Oberstein*), consists of two uprights (S), with a horizontal axis carrying a spring (F)—which vibrates 100 D.V. per second—with a writing-style at its free end (Fig. 120). A brass plate (B—*b*) moves in a slot, and carries a

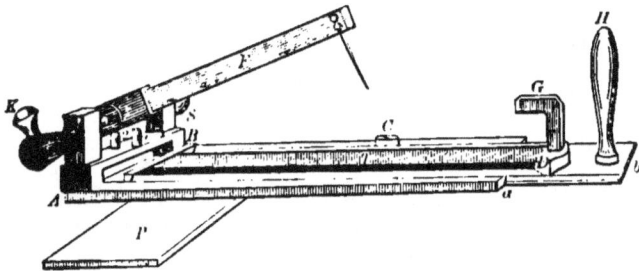

Fig. 120.

smoked glass plate (T), a catch (DG), and a handle (H). The handle (H) pushes up the glass plate and catch (G) until the latter meets the spring (F), and puts (F) on the stretch. When the catch (G) is withdrawn, (F) vibrates, and if the style be arranged to touch the glass, a curve is obtained on the latter.

(a.) It requires two persons. The observed person places a finger on the knob (K), while the catch (G) and glass plate are pushed up, the former to catch on (F), and the style is arranged to write on the glass. The observed person must not look, but close his eyes and listen.

(b.) The observer suddenly pulls on (H), thus discharging the spring (F), which vibrates and produces a note. The moment the observed person hears the sound, he presses the knob (K) and raises the writing-style. Of course, a curve is recorded, and it is easy to calculate the time which has elapsed between the emission of the sound and the reaction by the observed person. Numerous observations must be made, and the mean taken.

(c.) The instrument may also be used for vision—*i.e.*, when the slide (B—b) on being moved uncovers a painted disc.

(d.) In the more complete form of the apparatus, a key is fixed on one side of the apparatus, so that an electrical current is made or broken at the moment the spring begins to vibrate. The key is placed in the primary circuit of the induction machine, and the electrodes of the secondary battery are applied to any part of the skin, the observed person depressing the knob (K) when he feels the stimulus. One can thus make numerous experiments on the "Reaction Time" from different parts of the body.

3. **Another Method—Apparatus.**—Two Grove's cells, Morse and du Bois key, Deprèz' signal, chronograph in circuit with a tuning fork of 250 D.V. per second, electro-magnet with a writing-style, drum moving about 30 cm. per second, induction machine, electrodes, wires.

(a.) Arrange in circuit the two Grove's cells, a du Bois key, induction coil for single shocks, Deprèz' signal with writing-style, and a Morse key. Let the du Bois key be near the observer, and the Morse key near the observed person. Arrange the signal to record on the drum, and directly under it allow a chronograph (250 D.V.) to record its vibrations.

(b.) Suppose the electrodes from the induction machine are applied to the tongue of the person to be experimented on, start with the du Bois key open, and the Morse key closed. The observer suddenly closes the du Bois key, the observed person being so placed that he cannot see when this is done. This moment is recorded by the signal as it is in circuit. The observed person feels the induction shock, and at the same moment he presses the Morse key, and thus breaks the circuit, whereby the armature of the signal is set free. The interval between the two events, calculated from the vibrations of the tuning-fork, gives the reaction time. The experiment must be repeated several times, and the mean of several observations taken.

4. Inhibition.

(*a*.) Take an uninjured frog, place it on its back, and observe that it will not lie in this position, but immediately rights itself. Tie pretty firmly a thick string round each upper arm. This in no way interferes with the movements of the frog, but on placing the animal on its back, it no longer rights itself, but continues to lie in this position for a long time. It may be moved or pulled by the legs, yet it does not regain its normal attitude. Notice the modification of the respiratory movements.

5. Kircher's Experimentum Mirabile.

(*a*.) Take a hen and hold it gently to restrain its movements. Bring its bill in contact with a table, and then with a piece of white chalk draw a line directly outwards from its bill. Hold the animal steadily for a few seconds, and on removing the hands gently, it will be found that the hen lies quiescent and does not move for a considerable time. It may be rolled to one side or the other, yet it lies quiescent.

(*b*.) Repeat the same process, but instead of a white line lay a straw or white thread over the base of its bill. In a short time the animal becomes quiescent. Notice the alteration of the heart-beat and the depth and number of the respirations.

PHYSIOLOGY OF THE SENSE ORGANS.

— — —

FORMATION OF IMAGE — DIFFUSION— ABERRATION — ACCOMMODATION — SCHEINER'S EXPERIMENT—NEAR AND FAR POINTS— PURKINJE'S IMAGES— PHAKOSCOPE—ASTIGMATISM.

1. **Formation of an Inverted Image on the Retina.**

(*a.*) Fix the fresh excised ox-eye provided for you, remove the sclerotic from part of the posterior segment of the eye near the optic nerve. Roll up a piece of blackened paper in the form of a tube, black surface innermost, and place the eye in it with the cornea directed forwards. Look at an object—*e.g.*, a candle-flame—and observe the *inverted* image shining through the retina and choroid.

(*b.*) Fix the fresh excised eye of an albino rabbit in du Bois-Reymond's apparatus provided for you, and observe the same phenomenon. The eye is fixed in with moist modeller's clay. Observe the effect on the retinal image when a convex or concave lens is placed in front of the cornea. These lenses rotate in front of the cornea, and are attached to the instrument.

(*c.*) Focus a candle-flame or other object on the ground glass plate of an ordinary camera for photographic purposes, and observe the small inverted image.

2. **Diffusion.**

(*a.*) In a dark room place a lighted candle or gas burner conveniently, and by means of a convex lens focus the

image of the flame on a sheet of white paper. It is better
to introduce a blackened cardboard screen with a narrow
hole in it between the light and the lens. Observe that a
sharp image is obtained only at a certain distance from the
lens. If the white screen be nearer or further away, the
image is blurred.

(*b.*) Fix a long needle in a piece of wood, or use a pencil
or penholder, close one eye, and bring the needle or pencil
gradually nearer to the other eye. After a time, when the
needle is five to six inches distant, it will no longer be
distinct, but blurred, dim, and larger.

(*c.*) Prick a smooth hole in a card with a needle, arrange
the needle at the proper distance to obtain the previous
diffusion effect, and now introduce the card between the
needle and the eye, bringing the card near the eye, and
looking through the hole in the card. The needle will
appear distinct and larger; it is distinct because the diffusion circles are cut off, and larger because the object is
nearer the eye.

3. Spherical Aberration.

(*a.*) Make with a needle a hole in a blackened piece of
cardboard, look at a light placed at a greater distance than
the normal distance of accommodation. One will see a
radiate figure, with four to eight radii. The figures obtained from opposite eyes will probably differ in shape.

4. Chromatic Aberration.   Coloured Fringes.

(*a.*) Fix steadily the limit between a white and black
surface, and while doing so bring an opaque card between
one eye and the object (the other eye being closed), with its
edge parallel to the limit between the white and black
surfaces, so as to cover the larger part of the pupil. The
margin next the black appears with a yellow fringe when
the part of the pupil which lies next the black surface is
covered, while there is a blue fringe in the opposite condition.

(*b.*) Look at a candle or gas-flame through a small hole
in a black card with cobalt glass placed behind the hole,

which allows only the red and violet rays to pass through it. Accommodate for the violet rays or approach the light, the flame appears violet, surrounded with a reddish halo; on accommodating for the red, or on receding, the centre is reddish with a violet halo.

(c.) **V. Bezold's Experiment.**—Make a series (10 to 12) of concentric circles, black and white alternately, each 1 mm. thick, the diameter of the whole being about 15 mm. On looking at these circles when they are placed within the focal distance, one sees the white become pink; to some eyes it appears yellow or greenish. The same is seen on looking at concentric black and white circles, or parallel black and white lines from a distance outside the far point of vision, the white appears red and the black bluish.

5. Accommodation.

(a.) Standing near a source of light, close one eye, hold up both forefingers not quite in a line, keeping one finger about six or seven inches from the other eye, and the other forefinger about sixteen to eighteen inches from the eye. Look at the *near* finger, a distinct image is obtained while the far one is blurred or indistinct. Look at the *far* image, it becomes distinct, while the near one becomes blurred. Observe that in accommodating for the near object one is conscious of a distinct effort.

(b.) Ask some one to note the diameter of your **pupil** when you accommodate for the near and distant object respectively. In the former case the pupil contracts, in the latter it dilates. Ask a person to accommodate for a distant object, and look at his eye from the side and somewhat from behind, the half of the pupil projects beyond the margin of the cornea. When he looks at a near object in the same line, and without moving the eyeball, observe that the whole pupil and a part of the iris next the observer are projected forwards, owing to the increased curvature of the anterior surface of the lens.

(c.) Hold a thin wooden rod or pencil about a foot from the eyes, and look at a distant object. Note that the object appears double. Close the right eye, the left image disappears and *vice versâ.*

(*d.*) At a distance of six inches from the eyes, hold a veil or thin gauze in front of some printed matter placed at a distance of two feet or thereby. Close one eye, and with the other, one soon sees either the letters distinctly or the fine threads of the veil, but one cannot see both equally distinct at the same time. The eye, therefore, can form a distinct image of a near or distant object, but not of both at the same time, hence the necessity for accommodation.

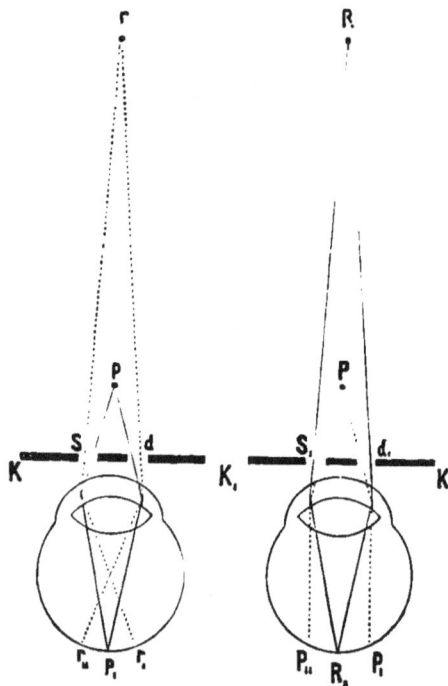

Fig. 121.—Scheiner's Experiment.

6. Scheiner's Experiment.

(*a.*) Prick two smooth holes in a card at a distance from each other less than the diameter of the pupil. Fix two long fine needles or straws in two pieces of wood or cork. Fix the cardboard in a piece of wood with a groove made in it with a fine saw, and see that the holes are horizontal. Place the needles in line with the holes, the one about eight inches and the other about eighteen inches from the card.

(*b.*) Close one eye, and with the other look through the holes at the *near* needle, which will be seen distinctly, while the far needle will be double, but both images are somewhat dim (Fig. 121).

(*c.*) With another card, while accommodating for the *near* needle, close the right-hand hole, the right-hand image disappears ; and if the left-hand hole be closed, the left-hand image disappears.

(*d.*) Accommodate for the *far* needle, the near needle appears double. Now close the right-hand hole, and the left-hand image disappears ; and on closing the left-hand hole, the right-hand image disappears (Fig. 121).

(*e.*) Instead of using a card perforated with two holes, use the apparatus provided for you, so constructed that one hole is covered with a green and the other with a red glass. Repeat the previous observations, noting the disappearance of the red or green image, as the case may be.

(*f.*) If desired, the holes in the card may be made one above the other, but in this case the pin looked at must be horizontal.

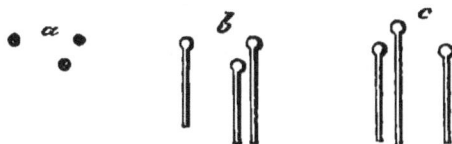

Fig. 122.

(*g.*) Make three holes in a piece of cardboard, as in Fig. 122, so that they can be brought simultaneously before one eye, and look at a pin or needle. One sees three images of the needle. On looking at a near object, the needles are in the position *b*, and at a distant object in that shown in *c*.

7. Determination of Near and Far Points.

(*a.*) Place one of the vertical needles used in the previous experiment conveniently, and with one eye—the other eye being closed—look through the two holes in a card, and when one distinct image of the needle is seen, gradually approximate the needle to the cardboard; observe that it becomes double, at a certain distance from the eye. This indicates the *near point of accommodation*.

(*b.*) Holding the card in front of one eye, gradually walk backwards while looking at the needle, observing when it becomes double. This indicates the *far point of accommodation*. *N.B.*—The experiment (*b.*) succeeds best in short-sighted individuals.

(*c.*) Determine the near point with a vertical needle and card with horizontal holes, and again with a horizontal needle with a card with the holes vertical. The two measurements do not usually coincide, because the curvature of the cornea is usually different in the two meridians.

8. Purkinje-Sanson's Images.

(*a.*) In a dark room, light a candle, and hold it to one side of the observed eye and on a level with it. Ask the person to accommodate for a distant object, and look into his eye from the side opposite to the candle, and three reflected images will be seen. At the margin of the pupil, and superficially, one sees a small bright *erect* image of the candle flame reflected from the *anterior surface of the cornea*. In the middle of the pupil there is a second less brilliant and not sharply defined *erect* image, which, of all the three images, appears to lie most posteriorly. It is reflected from the *anterior surface of the lens*. The third image lies towards the opposite margin of the pupil, is the smallest of the three, and is a sharp *inverted* image, from the *posterior surface of the lens*. Ask the person to accommodate for a near object, and observe that the pupil contracts, and the middle image —that from the anterior surface of the lens—becomes smaller and comes nearer to the corneal image. This shows that the anterior surface of the lens undergoes a change in its curvature during accommodation.

(*b.*) Instead of using a candle flame, cut two small square holes (10 mm. square) in a piece of cardboard, and behind each place a gas-flame, and observe the three pairs of square reflected images.

(c.) **Physical Experiment.**—Place in a convenient position on a table a large convex lens, supported on a stand. Standing in front of it, hold a watch-glass in the left hand in front of the lens and a few inches from it. Move a lighted candle at the side of this arrangement, and observe the three images described above. Substitute a convex lens of shorter focus, and observe how the images reflected from the lens become smaller.

9. The **Phakoscope of Helmholtz** is used to demonstrate more clearly the change in the curvature of the crystalline lens during accommodation (Fig. 123).

(a.) Place the phakoscope in a convenient position, darken the room. Two persons are required. The observed eye looks through a hole in the box opposite to c, while the observer looks through the hole (a) at the side. Light a lamp, place it some distance from the two prisms (b, b') in such a position that its light is thrown clearly upon the observed eye, and the observer sees two small bright square images of light, when the observed eye looks straight ahead at a distant object. These are the corneal images. He should also see in the observed eye, two larger less distinct images, from the anterior surface of the lens, and two smaller and much dimmer images, from the posterior surface of the lens. The last are seen with difficulty.

Fig. 123.—Phakoscope.—a, Hole for observer's eye; b, b', prisms; c, carries a needle for the observed eye to fix as its near point.

(b.) Ask the observer to accommodate for a near object—viz., the pin above c, keeping the eye unmoved. Observe that the middle image becomes smaller and goes nearer to the corneal one, while the other two undergo no perceptible change. At the same time the pupil becomes smaller.

10. Line of Accommodation—*i.e.*, the eye does not accommodate for a point, but for a series of points, all of which are equally sharply perceived with a certain accommodation.

(*a.*) Stretch a white thread about a metre long on a blackened wooden board. Through two narrow slits in a blackened card, about 2 mm. apart, focus with one eye a particular part of the thread, which must be in the optic axis. A part of the thread on the far and near side of the point focussed is quite distinct and linear, but beyond or nearer than this the thread is double and diverges from the point focussed.

(*b.*) Make a small black spot with ink on a glass plate and hold it in front of any printed matter. Bring the eye as close as possible to the glass plate without losing distinct definition of the point, we can see at one and the same time only one of the objects; but not the point, and the print equally sharply defined. Remove the eye gradually from the glass plate, and ultimately at a certain distance both the point and print will be equally distinct, and the point and print mark the extreme limits of the line of accommodation.

11. Astigmatism is usually due to unequal curvatures of the cornea in different meridians.

(*a.*) Draw on the card supplied to you two black lines of equal thickness, intersecting each other at right angles. Fix it vertically at the far limit of accommodation and look at it, when probably either the vertical or the horizontal line will be seen more distinctly. Test each eye separately. The line most distinct corresponds to the meridian of least curvature of the cornea.

(*b.*) Instead of a cross construct a star, the lines radiating at equal angles from the centre, and being of equal thickness. Repeat the previous observations, observing in which meridian the lines are most distinct.

(*c.*) Repeat these observations with the " astigmatic

clock " suspended on the wall, or with appropriate illustrations given in *Snellen's " Test-types."*

(*d.*) Construct a series of concentric circles of equal thickness and tint, about one-eighth of an inch apart upon a card.
Make a small hole in the centre of the card. Look steadily
at the centre of the card held at some distance. All the
parts will not be equally distinct. Approach the card
towards you, noting in which diameter the lines appear most
distinct.

(*e.*) This card may be used in another way. Hold the
card in front of and with the circles directed towards the
eye of another person—especially one with astigmatism—
place your own behind the hole in the card and look into
the observed eye, noting the reflection of the circles to be
seen in the eye. Observe in which meridian the circles are
most distinct, and if there be any perceptible difference in
the thickness and distinctness of the circles.

(*f.*) Draw a series of parallel, vertical, and horizontal lines
of equal tint and thickness, and about one-eighth of an inch
apart. Fix the card vertically at a distance, and move towards it, noting whether the vertical or horizontal lines are
most distinct.

(*g.*) Fix a fine wire or needle vertically in a piece of wood
moving in a slot, and similarly fix another needle or wire
horizontally. Move the needles until both can be seen
distinctly at the same time, when it will be found that the
needles are some distance apart, usually the horizontal one is
the nearer.

12. Diplopia Monophthalmica.

(*a.*) Make a small hole in a black card, hold it at some
distance, and with one eye look at a luminous point, the
eye being accommodated for a distant object. One sees
either several objects (feeble light) or an irregular radiate
figure with four or eight rays. Move the paper and the
long rays remain in the same position. Compare the figure
obtained from the other eye. It will very likely be different.

LESSON LXI.

BLIND SPOT — FOVEA CENTRALIS DIRECT VISION—MAXWELL'S EXPERIMENT — PHOSPHENES — RETINAL SHADOWS.

1. The Blind Spot.

(*a.*) **Marriotte's Experiment.**—On a white card make in the same horizontal line a black cross and a circle about three inches apart, as in Fig. 124. The cross and the circle may

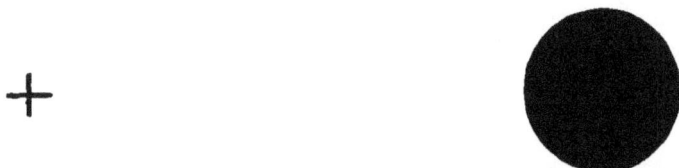

Fig. 124.—Marriotte's Experiment.

be black on white or coloured card. Hold the card vertically about ten inches from the right eye, the left being closed. Look steadily at the cross with the right eye, when both the cross and the circle will be seen. Gradually approach the card towards the eye, keeping the axis of vision fixed on the cross. At a certain distance the circle will disappear—*i.e.*, when its image falls on the entrance of the optic nerve. On bringing the card nearer, the circle reappears, the cross of course being visible all the time.

(*b.*) Perform the experiment in this way. Place the flat hand vertical to the face, and with its edge touching the

Fig. 125.

nose so as to form a septum between the two fields of vision. Fix the cross in Fig. 125, keep both eyes open, and on moving the paper to and fro at a certain distance both black dots will disappear.

(c.) Close one eye, and fix the point a (Fig. 126); on moving the paper a certain distance (about 16 cm.), one sees a complete cross, and to most observers the horizontal bar appears uppermost.

a
+

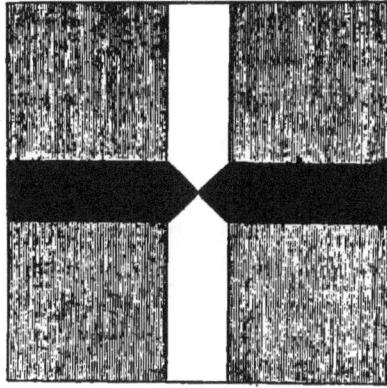

Fig. 126.

2. Map out the Blind Spot.

(a.) Make a cross on the centre of a sheet of white paper, and place it on a table about ten or twelve inches from you. Close the left eye, and look steadily at the cross with the right eye. Wrap a penholder in white paper, leaving only the tip of the pen-point projecting, dip the latter in ink, or dip the point of a white feather in ink, and keeping the head steady and the axis of vision fixed, place the pen-point near the cross, and gradually move it to the right until the black becomes invisible. Mark this spot. Carry the blackened point still further outwards until it becomes visible again. Mark this outer limit. These two points give the outer and inner limits of the blind spot. Begin again, moving the pencil first in an upward and then in a downward direction, in each case marking where the pencil becomes invisible. If this be done in several diameters, an outline of the blind spot is obtained, even little prominences showing the retinal vessels being indicated.

18

3. To Calculate the Size of the Blind Spot.

(*a.*) Helmholtz gives the following formula for this purpose :—When f is the distance of the eye from the paper, F the distance of the second nodal point from the retina—usually 15 mm.—d the diameter of the sketch of the blind spot drawn on the paper, and D the corresponding size of the blind spot:—

$$\frac{f}{F} = \frac{d}{D}$$

4. Acuity of Vision of the Fovea Centralis.

(*a.*) On a horizontal plane—a blackboard—describe a semicircle with a radius equal to that of the near point of vision, and fix in the semicircle pins at an angular distance of 5° apart. Close one eye, and with the other look at the central pin; the pins on each side will be seen distinctly, those at 10° begin to be indistinct, while those at 30° to 40° are not seen at all.

(*b.*) At a distance of five feet look at a series of vertical parallel lines alternately black and white, each ·5 mm. wide. A normal eye will distinguish them; if not, approach the object until they are seen distinctly.

5. Direct Vision.—When the image of an object falls on the fovea centralis, we have what is called "direct vision," but when the image falls on any other part of the retina, it is called "indirect vision." Vision is most acute at the fovea centralis of the yellow spot.

(*a.*) Standing about two feet from a wall, hold up a pen at arm's length between you and the wall. Look steadily at a fixed spot on the wall, seeing the pen distinctly all the time. Move the pen gradually to one side; first one fails to see the hole in the nib, and as the pen is carried outwards one fails to recognise it as a pen. Hence, in looking at a large surface, to see it distinctly one must unconsciously move his eyeballs over the surface to get a distinct impression thereof.

(*b.*) Make two black dots on a card quite close together so that when looked at they are seen as two. Hold up the left index finger, look steadily at it, and place the card with the dots beside the finger. Move the card outwards, inwards, upwards, and downwards successively, and note that as the dots are moved towards the periphery they appear as one, but not at equal distances from the fixed point in all meridians. For convenience, the card may be moved along a rod, movable on a vertical support.

6. Maxwell's Experiment. The Yellow Spot.

(a.) Make a strong watery solution of chrome alum—filter it, and place it in a clear glass bottle with flat sides. Close the eyes for a minute or so, open them, and while holding the chrome alum solution between one eye and a white cloud, look through the solution. An oval or round rose spot will be seen in the otherwise green field of vision. The pigment in the yellow spot absorbs the blue-green rays, hence the remaining rays which pass through the chrome alum give a rose colour.

7. Bergmann's Experiment.

Make a series of parallel vertical black lines, 2 mm. in diameter, on white paper, with equal white areas intervening between them. Look at them in a good light, at a distance of two to three yards. In a short time the lines will appear as in Fig. 127, A. They appear of this shape because of the manner in which the images of the lines fall on the cones in the yellow spot, as shown in B.

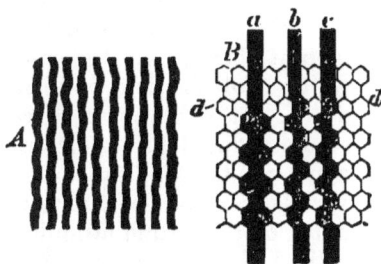

Fig. 127.—Bergmann's Experiment.

8. Phosphenes.

(a.) Press the finger firmly, or better still the head of a large pin, against the inner corner of the closed eye. A brilliant circular patch, with a steel-grey centre and yellow circumference, is seen in the *field of vision* and on the opposite side, and of the same shape as the compressing body.

Press any other part of the eyeball, the same spectrum is seen, and always on the opposite side. Impressions made on the terminations of the optic nerve are referred outside the eye—*i.e.*, beyond into space. The phosphene is seen in the upper half if the lower is pressed, and *vice versâ*.

9. Shadows of the Fovea Centralis and Retinal Blood-vessels.

(*a.*) Move, with a circular motion, a blackened card, with a pin-hole in its centre, in front of one eye looking through the pin-hole at a white cloud. Soon a punctated field appears with the outlines of the capillaries of the retina. The oval shape of the yellow spot is also seen, and it will be noticed that the blood-vessels do not enter the *fovea centralis*. Move the card vertically, when the horizontal vessels are most distinct. On moving it horizontally, the vertical ones are most distinct. Some observers recommend that a slip of blue glass be held behind the hole in the opaque card, but this is unnecessary.

10. Purkinje's Figures.

(*a.*) Darken a room, light a candle, and stand in front of a monochromatic wall. If this is not available, hang up a large white sheet, and while looking steadily with one eye towards the wall or sheet, accommodating the eye for a distant object, hold the candle close to the side of that eye, well out of the field of vision, and move the candle up and down. It is better to direct the eye outwards, keeping it accommodated for a distant object. Ere long, dark somewhat red-brown branching lines, shadows of the retinal vessels, will be seen on a dark background. Therefore the parts of the retina stimulated by light must lie behind the retinal blood-vessels.

11. Muscæ Volitantes.

(*a.*) Light a candle in a dark room; at a distance from it place a black screen with a pin-hole in it. Focus by means of a convex lens the image of the flame upon the hole in the screen. Look through the hole with one eye, and on the illuminated part of the lens will be seen images of dots and threads due to objects within the eyeball.

12. Inversion of Shadows thrown on the Retina.

(*a.*) Make three pin-holes in a card, and arrange them in a triangle close to each other. Hold the card four or five inches from the right eye, and look through the holes at a bright sky or lamp. Close the left eye, and in front of the right hold a pin so that it just touches the eyelashes. An inverted image of the pin will be seen in each pin-hole. Retinal images, as we have seen, are inverted on the retina, shadows on the retina are erect, and therefore the latter on being projected outwards into space are seen inverted.

13. Duration of Impressions.

(*a.*) On a circular white disc, about half way between the centre and circumference, fix a small black oblong disc, and rapidly rotate it by means of a rotating wheel. There appears a ring of grey on the black, showing that the impression on the retina lasts a certain time.

14. Talbot's Law.—A grey once produced is not changed by

increased rapidity of rotation of the disc exciting the sensation. The intensity of the light impression is quite independent of the absolute duration of the periods of illumination and shade.

(*a.*) Rotate a disc like Fig. 128 twenty-five times per second, then the period in which illumination and shade alternately lasts for the inner zone is $\frac{1}{25}$ sec., for the middle $\frac{1}{50}$, and for the outer zone $\frac{1}{100}$ sec. In

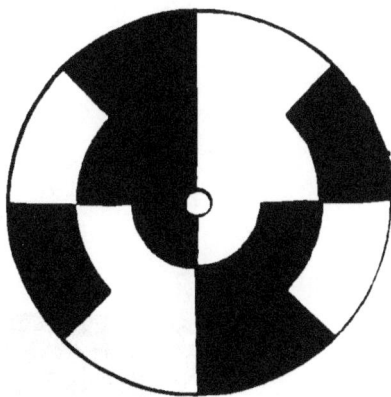

Fig. 128.

all three zones, the period of illumination lasts exactly one-half of the period, and the three zones have exactly the same brightness. Rotate more quickly, and no further effect is produced. The number of rotations is readily determined by Harding's improved counter.

LESSON LXII.

PERIMETRY—IRRADIATION—IMPERFECT VISUAL JUDGMENTS.

1. To Map out the Field of Vision, or Perimetry.

(*a.*) A rough method is to place the person with his back to a window, ask him to close one eye, stand in front of him about two feet distant, hold up the forefingers of both hands in front of and in the plane of your own face. Ask the person to look steadily at your nose, and as he does so,

Fig. 129.—Priestley Smith's Perimeter.

observe to what extent the fingers can be separated horizontally, vertically, and in oblique directions before they disappear from his field of vision.

(*b.*) **Priestley Smith's Perimeter** (Fig. 129).—Let the observer seat himself near a table on which the perimeter is placed at a convenient height. Suppose the right eye is to be examined, fix a blank chart for the right eye behind the wooden circular disc. A mark on the hand-wheel shows which way the chart is to be placed.

(*c.*) The patient rests his right cheek against the knob on the wooden pillar in such a position that the knob is about an inch directly under his right eye. The other eye is closed either voluntarily or with a shade, while the observer looks steadily with the right eye at the white spot on the end of the axis of the instrument.

(*d.*) The observer turns the quadrant with his right hand, by means of the wooden wheel, first to one and then to another meridian. With his left he moves the white mark along the quadrant, beginning at the periphery and gradually approaching centralwards until it is just seen by the right eye. A prick is then made in the chart corresponding to the angle read off on the quadrant, at which the observer can see the white spot.

(*e.*) Turn the quadrant to another meridian and determine the limit of the visual field as before. This is repeated for four or more meridians, and then the pricks on the chart are joined by a continuous line, when we obtain an oval field more extensive in the outer and lower portions. Test, if desired, the left eye, substituting a blank chart for that eye.

(*f.*) Test the field of vision for colours, substituting for the white travelling disc, blue, red, and green. Mark each colour-field on the chart with a pencil of similar colour. Notice that the field for blue is nearly as large as the normal visual field. It is smallest for green, red being intermediate between green and blue.

(*g.*) With Ludwig's apparatus test when red, yellow, blue, and other coloured glasses cease to be distinguished as such in the field of vision.

2. Irradiation.—By irradiation is meant the fact that, under certain circumstances, objects appear larger than they should be according to their absolute size and distance from the eye, larger than other objects of greater or less brightness of the same size and at the same distance.

(*a*). Cut out two circles as in Fig. 130, or two squares of

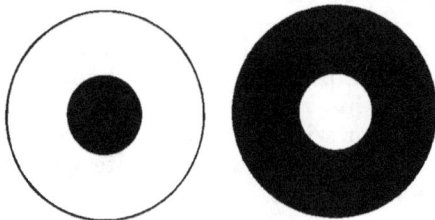

exactly the same size, of white and of black paper. Place the white patch on a black, and the black on a white sheet of paper. Hold them some distance from the eye, and especially if they be not

Fig. 130.—Irradiation.

distinctly focussed, the white circle will appear larger than the black one.

(*b*.) Divide a square into four as shown in Fig. 131, two of the smaller squares being white and two black. Hold the

Fig. 131.

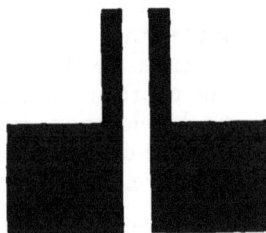

Fig. 132.

figure at some distance from you. The two white squares appear larger, and they appear to run into each other and to be joined together by a white bridge.

(*c*.) Look at Fig. 132, placed at such a distance that the accommodation is imperfect. The white stripe which is of equal breadth throughout appears wedge-shape, being wider below between the broad black patches, and narrow above. To me also the narrow black patches appear to be broader above and narrower below.

(*d.*) Gum on to a sheet of white paper two strips of black paper 5 mm. wide, and parallel to each other, leaving a white interspace of 8 mm. between them. Look at the object, and, especially if it be not sharply focussed, the smaller black strips will appear broader than the white one.

3. Imperfect Visual Judgments.

(*a.*) Make three round black dots, A, B, C, of the same size, in the same line, and let A and C be equidistant from B. Between A and B make several more dots of the same size. A and B will then appear to be further apart than B and C.

(*b.*) Make on a white card two squares of equal size, omitting the outlines. Across the one draw *horizontal* lines at equal distances, and in the other make similar *vertical* lines. Hold them at some distance. The one with horizontal lines appears higher than it really is, while the one with vertical lines appears broader—*i.e.*, both appear oblong.

(*c.*) Look at the row of letters (S) and figures (8). To some the upper halves of the letters and figures may appear

S S S S S S S   8 8 8 8 8 8 8 8
Fig. 133.

to be the same size as the lower halves, to others the lower halves may appear larger. Hold the figure upside down, and observe that there is a considerable difference between the two, the lower half being considerably larger.

(*d.*) **Zöllner's Lines.**—Make two lines parallel to each other. Note that one can judge very accurately as to their parallelism. Draw short oblique lines through them. The lines now no longer appear to be parallel, but seem to slope inwards or outwards, according to the direction of the oblique lines.

(*e.*) Look at Fig. 134, the long oblique lines do not appear to be parallel although they are so.

Fig. 134.—Zöllner's Lines.

4. Imperfect Judgment of Distance.

(*a.*) Close one eye, and hold the left forefinger vertically in front of the other eye, and try to strike it with the right forefinger. On the first trial one will probably fall short of the mark, and fail to touch it. Close one eye, and rapidly try to dip a pen into an inkstand, or put a finger into the mouth of a bottle placed at a convenient distance. In both cases one will not succeed at first. In these cases one loses the impressions produced by the convergence of the optic axes, which are important factors in judging of distance.

(*b.*) Hold a pencil vertically about 15 cm. from the nose, fix it with both eyes, close the left eye, and then hold the

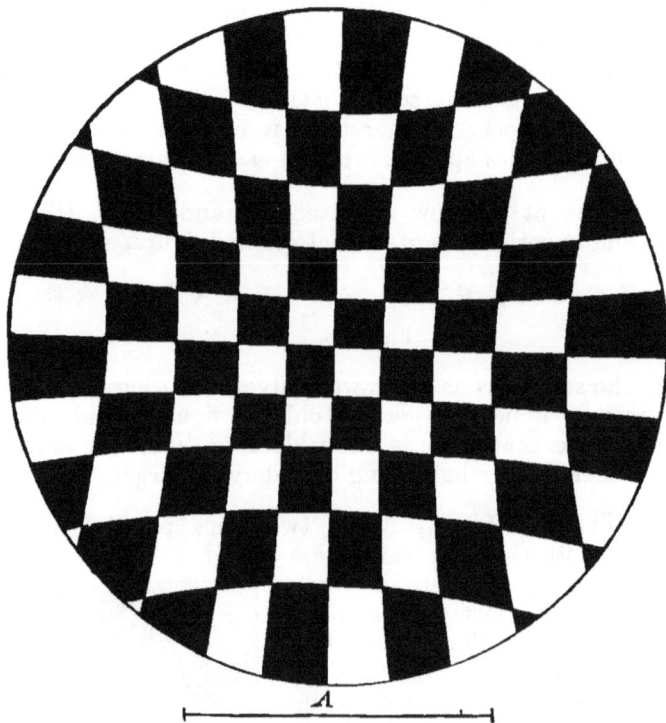

Fig. 135.

right index finger vertically, so as to cover the lower part of the pencil. With a sudden move try to strike the pencil with the finger. In every case one misses the pencil and sweeps to the right of it.

5. Perception of Size.

(*a.*) Fix the centre of Fig. 135 at a distance of 3 to 4 cm. from the eye, when by indirect vision the broad white and black areas of the peripheral parts, bounded by hyperbolic curves, will appear as small and the lines bounding them as straight as the smaller areas in the middle zone.

6. Apparent Movements.

(*a.*) Strobic Discs.—Give the discs a somewhat circular but rapid movement, and observe that the rings appear to move, each one on its own axis.

(*b.*) Radial Movement.—While another person rotates a disc like Fig. 136 on the rotating wheel, look steadily at the

Fig. 136.

centre of the disc. One has the impression as if the disc were covered with circles which, arising in the centre, and, gradually becoming larger, disappear at the periphery. After long fixation look at printed matter or at a person's face, the letters appear to move towards the centre, while the person's face appears to become smaller and recede. If the disc be rotated in the opposite direction, the opposite results are obtained.

LESSON LXIII.

KÜHNE'S ARTIFICIAL EYE—MIXING COLOUR SENSATIONS—COLOUR-BLINDNESS.

1. Kühne's Artificial Eye. (Fig. 137.)

(*a.*) Fill the instrument with water, and place it in a darkened room with the cornea directed to a hole in a shutter, through which sunlight is directed by means of a heliostat. If this is not available, use an oxy-hydrogen lamp or electric light to throw parallel rays of light on the cornea. If these cannot be had, use a fan-tailed gas burner, but in this case the illumination and images will be very feeble. To enable one to observe the course of the rays of light, pour some eosin or fluorescin into the water in the instrument.

(*b.*) Formation of the image on the retina. Observe the course of the rays of light, which come to a focus behind the lens—the principal posterior focus. Move the ground glass representing the retina, and get a clear inverted image of the source of light. *N.B.*—In this instrument accommodation is effected not by altering the curvature of the lens, as in the normal eye, but by moving the retina.

(*c.*) Place convex and concave lenses between the source of light and the cornea; observe how each alters the course of the rays and their focus.

(*d.*) After having an image well-focussed upon the retina, move the latter away from the lens, when the image becomes blurred owing to diffusion. If, however, a slip of zinc, with a hole cut in it to act as a diaphragm to cut off some of the marginal rays, be interposed, the image is somewhat improved.

(*e.*) After seeing that the light is sharply focussed on the retina, remove the lens—to imitate cataract—and observe that the rays are focussed quite behind the retina.

(*f.*) Place the removed lens in front of the cornea, the principal focus is now much in front of the retina, so that a much weaker lens than the one removed has to be used after removal of the lens for cataract.

(*g.*) **Astigmatism.**—Fill the plano-convex glass (*g*)—to imitate a cylindrical lens—with water, and place it in front of the cornea. Between the cornea and the cylindrical lens place a sheet of zinc with a cross cut out in it, or with a number of holes in a horizontal line. One cannot obtain a distinct image of the cross or the holes, as the case may be.

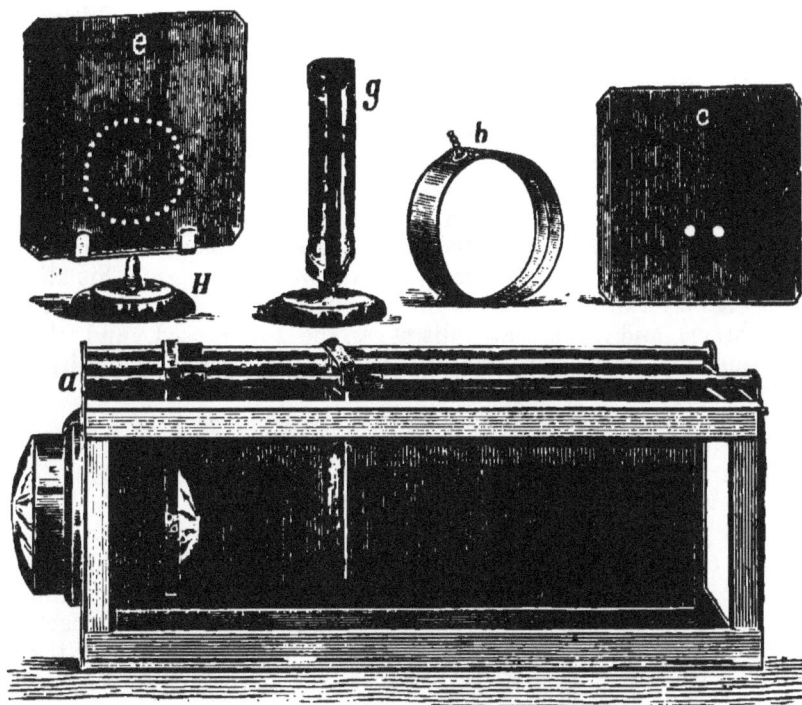

Fig. 137.—Kühne's Artificial Eye, as made by Jung of Heidelberg.

(*h.*) **Scheiner's Experiment.**—With the light properly adjusted, place in front of the cornea a piece of zinc perforated with two holes (*c*), 1 cm. in diameter, in a horizontal line, the distance between the holes being less than the diameter of the pupil. Find the position of the retina—and there is only one position—in which the two beams of light are brought to a focus. Move the retina towards the cornea, and observe two images; close the right-hand hole and the right-hand image disappears. Bring the retina posterior to the principal focus, and again there are two images. On closing the right-hand hole, the left-hand image disappears, and *vice versâ*.

2. Mixing of Colour Sensations.

(*a.*) Arrange on the spindle of the rotating apparatus the disc with coloured sectors provided for you. On rotating the disc rapidly, observe that it appears grey or whitish. The disc is provided with sectors corresponding to the colours of the spectrum, and arranged in varying proportions.

(*b.*) Arrange three of Maxwell's colour discs—red, green, and violet—upon the spindle of the rotating apparatus. Adjust the relative amounts of these three colours, so that on rapidly rotating them they give rise to the sensation of grey or white. Each disc is of a special colour, and has a radial slit from the centre to the circumference. This slit enables a disc of a different colour to be slipped over the other, and thus many discs can be superposed, and the amount of each colour exposed regulated in any desired proportion.

(*c.*) Combine a chrome yellow disc and a blue one in various proportions, and on rotating, the resultant colour is never green, but a yellowish or reddish grey.

(*d.*) Arrange two coloured discs of vermilion and bluish-green in the proportion of 36 of the former to 64 of the latter. On the same spindle arrange a white and a black disc—with a diameter a little more than half that of the former pair—the white being in the proportion of 21·3 to to 78·7 of the black. On rotating, a grey colour is obtained from both sets of discs.

(*e.*) **Lambert's Method.**—On a black background place a blue wafer or square of blue paper, and six or seven inches behind it a yellow square or wafer. Hold a plate of clear glass vertically, about ten inches above and midway between the two squares. Look obliquely through the glass, and get the reflected image of the yellow to overlap the blue, seen directly through the glass ; where they overlap appears *white.*

3. To Test Colour-Blindness.—On no account is the person being tested to be asked to name a colour. In a large class of students one is pretty sure to find one or more who are more or less colour-blind. The common defects are for red and green.

(*a.*) Place **Holmgren's worsteds** on a white background in a good light. Select, as a test colour, a skein of a green colour, such as would be obtained by mixing a pure green with white. Ask the examinee to select and pick out from the heap all those skeins which appear to him to be of the same colour, whether of lighter or darker shades. A colour-blind person will select amongst others some of the confusion-colours — *e.g.,* pink, yellow. A coloured plate showing these should be hung up in the laboratory. Any one who selects all the greens and no confusion-colours, has normal colour 'vision. If, however, one or more confusion-colours be selected, proceed as follows :—Select, as a test colour, a skein of pale rose. If the person be red-blind, he will choose blue and violet; if green-blind, grey and green.

(*b.*) Select a bright red skein. The red-blind will select green and brown ; the green-blind picks out reds or lighter brown.

4. Successive Light Induction.

(*a.*) Look for one minute at a small white circular disc on a black background—*e.g.,* velvet. Close and cover the eyes. A negative after-image of the disc appears, but it is darker and blacker than the visual area, and it has a peculiar light area round it, brightest close to the disc, and fading away from it.

(*b.*) Look at two small white square patches of paper placed one-eighth of an inch apart on a black background. On closing the eyes, the black space between them looks brighter than the other three sides of the squares.

(*c.*) Look at a black strip on a white ground. On closing

the eyes there is no partial darkening of the white ground, but only an intensely bright image of the strip.

5. Contrast.

(*a.*) On the rotating machine cause a disc, as in Fig. 138, to rotate with moderate rapidity, when several zones will be seen, the innermost black, while each one further outwards

Fig. 138.—Disc for Contrast.

is lighter in tint. Each zone, where it abuts against the inner darker zone, is lighter than the rest of the same zone, and shades off gradually to the outer part of the zone.

(*b.*) Place four lighted candles in a dark room before a white surface, and push between the candles and the screen towards the centre of the series an opaque screen—*e.g.*, cardboard, with a clean cut vertical edge. A part of the white surface is illuminated by all four candles, then a vertical area illuminated by three, and so on, and finally a part not illuminated by any of the candles. Each of these areas is throughout its entire extent equally illuminated, yet on the side where each area abuts against a darker area it appears lighter, on the other side darker, and gradually shaded between its outer and inner limits. This is due to the fact that strong stimulation of one part of the retina diminishes the excitability in the other parts, and the parts most affected are those next the excited area. Thus a change in the excitability of one part of the retina is brought about by stimulation of an adjacent part.

(*c.*) Place a small white square or oblong piece of paper on a dull, dead, black surface. Stare steadily at the white square, and observe that the edges appear whiter than the centre.

(*d.*) Place side by side a white and black surface. Cut two oblong ($1\frac{1}{2}'' \times \frac{1}{2}''$) pieces of grey, yellow, or other coloured paper of exactly the same size, and lay one piece of the grey on the white background, and the other on the black. Observe how much brighter the latter looks owing to contrast. Reverse the pieces, and notice that the same result occurs. Repeat with other colours.

(*e.*) Place on a table a small sheet ($4'' \times 4''$) of red and one of green paper. Cut out of a sheet of red paper two pieces about one inch square, and place them on the two large squares. Observe that the small red square on the green ground appears far brighter and more saturated than the red square on the red ground.

(*f.*) Cut a small hole (5×5 mm.) in a piece of coloured paper—*e.g.*, red—and look through the hole at a sheet of white paper, the hole appears greenish.

(*g.*) On an ordinary mirror place a slip of transparent coloured glass—red or green, or any other colour. Hold in front of the coloured glass a narrow strip of white paper; by adjusting the position of the glass in relation to the light, we see two images reflected from the anterior and posterior surface of the mirror; one has the same colour as the coloured glass, while the other or posterior one has the complementary colour—if a red glass be used, the latter is green—if a green glass, it is red. Hold in front of the red glass a piece of white paper with black printed matter on it. The black print is seen green in the posterior image. Gum a few narrow strips of white paper (1 mm. in diameter) on black paper, and on holding it up in front of the red glass, as before, the anterior image appears in the complementary of the glass—viz., green.

6. Simultaneous Contrast.

(*a.*) Cut out a small oblong of white, or preferably of grey paper, and put it on a large piece of bright green paper (4 inches square); the grey suffers no change. Cover the whole with a thin semi-transparent sheet of tissue paper. The grey oblong appears *pink*.

(*b.*) Instead of green paper, place the grey slip on red, and cover it as before; a greenish-blue contrast colour is seen.

(*c.*) Repeat (*a.*), but place a red square on a grey ground; the red square will appear greenish.

(*d.*) A grey square upon blue appears yellow; a yellow upon blue appears white, when covered with tissue paper.

(*e.*) Surround the small square with a broad black line, each square appears in its own colour. The effect of contrast is destroyed.

(*f.*) Place side by side two strips of paper, green and red (6 × 3). Over the line of junction place a strip of grey paper (¼ × 6), and cover the whole with tissue paper, as before. The grey appears pink on the green side, and greenish on the red. This contrast is also set aside by running a black margin round the grey strip. Do the same with yellow and blue.

(*g.*) Arrange a disc like Fig. 139 on the rotating wheel. On a white disc fix four narrow, coloured (*e.g.*, green) sectors,

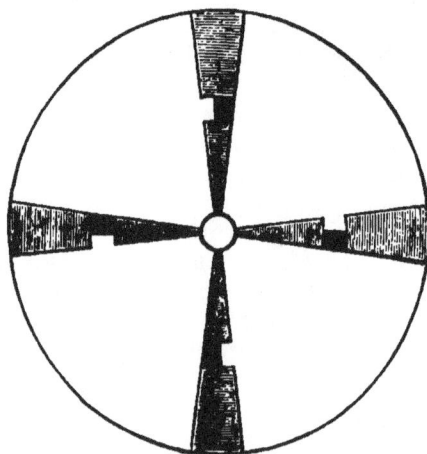

Fig. 139.

and interrupt each in the middle, as in the figure, with a black and white stripe. On rotating the disc, the ring which one might expect to be grey, from the black and white, appears reddish—*i.e.*, the complementary colour of the greenish ground.

(*h.*) Place a strip of grey paper on a black background and a corresponding strip on a white ground. The former will appear much lighter, the grey on white much darker. Fix the eyes for a minute on a point midway between the strips, close and cover the eyes. The after-images will show a great difference in luminosity.

(*i.*) **Ragona Scina's Experiment.**—Two pieces of wood fixed at right angles to each other are covered by white paper, while a coloured sheet of glass is held at an angle of 45° between them (Fig. 140). Look vertically through the glass at the horizontal white paper, and observe a pale red tint. Attach a small black square to the centre of the vertical arm at B, the image of this square is seen at *b* as a deep red image. Place a similar black square on the horizontal board at C, it should appear grey; but a grey on a red ground causes contrast, and so one sees a greenish-blue square alongside a red one.

Fig. 140.—Ragona Scina's Experiment.

7. Coloured Shadows.

(*a.*) Place an opaque vertical rod (1 inch in diam.) in front of a white background. Admit not too bright daylight to cast a shadow of the rod. Place a lighted candle behind one side of the rod, the shadow caused by the yellow-red light of a candle, and illuminated by the daylight, appears blue—*i.e.*, a purely subjective blue, the complementary colour of the yellow-red light of the candle, which casts a yellow light. The effect is more pronounced, the darker both shadows are. To show that the blue is purely subjective, roll up a sheet of black paper—black surface innermost—in the form of a tube about $\frac{1}{4}$ inch or less in diameter. At a distance of 18 inches, look at the centre of the blue shadow, and let an observer cut off the light from the candle by means of an opaque screen. On removing the screen, no change is visible, but if the tube be directed to the line of junction of the blue shadow, with the illuminated background just beyond it, the blue appears.

(*b.*) In a window-shutter of a dark room, cut two square holes (10 cm.) on the same horizontal plane, and 2 feet apart. In one fix a piece of clear glass to admit ordinary white light, and into the other fit a red or green coloured glass. Both openings must be provided with a movable shutter to regulate the amount of light admitted. At 3 to 4 feet distance, place a rod or flat piece of wood vertically against a white surface. Observe two shadows. Suppose the glass to be red, then the shadow due to the ordinary light is red, that of the red glass is greenish. Substitute for the red light that of a lighted candle. The shadow then appears blue.

8. Choroidal Illumination.

(*a.*) In a dark room, light an ordinary lamp or fan-tailed gas-burner. Place the source of light at the right side, about 2 feet from an open book or sheet of paper. Partly separate the fingers of the left hand and place them over the face, so that different portions of the paper are seen by each eye. That half of the page seen with the right eye has a *greenish* tint, the other part seen with the left eye is red or *pinkish*. Change the source of light to the left side, the colours are reversed.

(*b.*) With the conditions as in (*a.*), hold a piece of paper (3–4 cm. wide), or a visiting card, between the eyes with its flat surface towards the face, the same phenomena are seen.

(*c.*) Cut in a piece of black cardboard two rectangular holes (4 × 10 mm.), separated by a distance about equal to that between the pupils, with the conditions as in (*a.*) Hold the cardboard about 10 inches or more from you, and look through the holes at a white surface; four images of the two holes will be seen, the inner right and outer left images are impressions from the right eye, the inner left and outer right from the left eye. This is easily proved by closing either eye, when the images belonging to that eye disappear. If the source of light be on the right side, the former pair of images is greenish in colour, the latter is pale pink. Change the light to the left side and the colours are reversed (*H. Sewall*). The colour-phenomena occur without the aid of objective colour, and are due to light passing through the sclerotic and choroid coats.

9. Binocular Contrast.

(*a.*) Place a white strip of paper on a black surface, look at the white paper and squint so as to get a double image. In front of the right eye place a blue glass, and in front of the left one a grey (smoked) glass. The image of the right eye will be blue, that of the left yellow. Instead of the grey glass, a card with a small hole in it placed in front of the left eye does perfectly well. The yellow of the left eye is a contrast sensation.

10. Positive After-Images.

(*a.*) In a room not too brightly illuminated, rest the retina by closing the eyes for a minute or two, then suddenly look for a second or two at a gas jet surrounded with a white globe, then close the eyes. An image corresponding exactly to that looked at will be seen.

(*b.*) After resting the retina by closing the eyes, look at a gas flame surrounded with a coloured glass, or look at a gas flame in which some substance is burned to give a characteristic flame—*e.g.*, common salt. Then look at a white surface, when a positive after-image of the same colour will be seen. In all these cases the image moves as the eye is moved, showing that we have to do with a condition within the eye.

11. Negative After-Images.

(*a.*) Rest the retina, and then stare steadily for half a minute or less at a small white square or white cross on a dead black ground. To insure fixation of the eyeballs, make a small mark in the centre of the white paper, and fix this steadily. In all subsequent experiments do the same. Then suddenly slip a sheet of white paper over the whole, a *black* square or cross will appear on the white background. I find that the best black surface to use is the dull dead black of the "Tuch-papier," such as is used by opticians for lining optical apparatus. Notice also while staring at the white paper, that its margins appear much brighter than the centre, owing to contrast.

(*b.*) The black negative after-image may also be seen by closing the eyes.

(*c.*) Look at a dull, dead, black square or cross on a white ground. Turn to a grey surface, when a white square or cross will appear.

(*d.*) Stare intensely at a bright red square on a black surface for twenty seconds, and then look at a white surface, a bluish-green patch on the white is seen. It waxes and wanes, and finally vanishes.

(*e.*) A green stared at in the same way gives a red—*i.e.*, in each case the complementary colour is obtained as a "negative coloured after-image."

(*f.*) Place a small red and a green square side by side on a black background, stare at them, and quickly cover the whole with a sheet of white paper, a greenish-blue after-image will appear in place of the red, and a reddish-purple instead of the green.

12. **The Haploscope** consists of two tubes about an inch and a quarter in diameter and 8 inches in length, which can be converged or diverged from a fixed point as desired. On looking through the tubes with both eyes, each eye has its own special field of vision, but with the proper convergence the two fields are united, and form one field of vision.

(*a.*) With the discs provided for you—copies of those of Volkmann—study the combinations obtained in the single field of vision.

(*b.*) **Struggle of the Fields of Vision.**—Place in one tube a red and in the other a green glass. One sees either a red or green disc, but not a mixture of the two colours.

13. **Stereoscope.**

(*a.*) Examine a series of stereoscopic slides to show the combination of the images obtained by the right and left eyes respectively—a slide to show lustre, and another of different colours on the two sides.

LESSON LXIV.
THE OPHTHALMOSCOPE.

The Ophthalmoscope.—Two methods are employed, and the student must familiarise himself with both.

1. Direct—giving an upright image.

2. Indirect—giving an inverted image.

The observer must practise both methods on another person, or use a rabbit for the purpose, or use an artificial eye.

A. Human Eye. 1. The Direct Method.

(*a.*) About twenty minutes before the examination is commenced, instil a drop of solution of sulphate of atropia (2 grains to the ounce of water in a drop-bottle), or hom-atropin into, say, the *right* eye of a person with normal vision. The pupil is dilated and accommodation for near objects is paralysed, owing to the paralysis of the ciliary muscle. The patient is seated in a darkened room, and the observer seats himself in front of him, and on a slightly higher level. Place a brilliant light obscured everywhere except in front, on a level with the left eye of the patient.

(*b.*) The observer takes the ophthalmoscope mirror in the right hand, resting its upper edge upon his eyebrow, holds it in front of his own eye, looking through the central hole in it, and directs a beam of light into the observed eye, when a red glare—the reflex—is observed. The patient is told to look upwards and inwards, which is conveniently accomplished by telling him to look to the little finger of the operator's right hand. The operator then moves the mirror, with his eye still behind it, and looks through the hole until the mirror is within *two to three inches* from the observed eye, taking care all the time that the beam of light is kept steadily thrown into the eye. If the eyes of the observer and patient be normal, the observed has simply to relax his accommodation—*i.e*, look as it were at a distant object, when the retina comes into view as an erect or upright object.

(*c.*) Observe the *retinal blood-vessels* running in different directions on a red ground. Move the mirror about to find the *optic disc*, with the central artery emerging from it. Trace the course of the veins accompanying the arteries across the disc.

2. The Indirect Method, giving an inverted image.

(*a.*) The patient, the light, and the observer are as before. The observer places himself about twenty to eighteen inches from the patient, and, holding the mirror in his right hand, by means of the mirror he throws a beam of light into the eye of the patient. When the eye is illuminated, he takes a small biconvex lens of two to three inches focus in his unemployed hand—the left in this case—holding it between his thumb and index finger, placing it vertically two or three inches from the observed eye. To ensure that the lens is held steadily, rest the little finger upon the temple or forehead of the patient. Keep the lens steady, and move the mirror until the optic disc is seen, with the details already described.

Both methods ought to be practised, as each has its advantages. In the direct method only a small part of the retina is seen at one time, but it is considerably magnified ; while by the indirect method, although more of the retina is seen at once, it is magnified only slightly.

If the observed or observer's eye is abnormal, suitable glasses to be fixed behind the mirror are supplied with every ophthalmoscope. In some forms of ophthalmoscope, such as that of Gowers and others, these lenses (convex +, and concave −) are fixed to a rotating disc behind the mirror. As the disc is rotated, lens after lens can be brought to lie exactly behind the hole in the mirror, and thus correct any anomaly of refraction.

3. Eye of a Living Rabbit.

(*a.*) Instil atropin as before, or use an atropinised gelatin disc to effect the same result. Place the rabbit in a suitable cage to keep it from moving. A very suitable one was devised by Michel, use it. (Fig. 141.) Examine the eye by the direct and indirect methods already described. *N.B.*—If an albino rabbit be used, the observer sees the large choroidal vessels.

4. Perrin's Artificial Eye.

Fig. 141.—Carriage for Rabbit.

(*a.*) Use this until a clear image of the fundus is obtained by both methods. In fact, it is well for the student to begin with this. In this model eye-caps to fit on to the eye are supplied, so as to render the eye-model either myopic or hypermetropic. Afterwards test these, and use the necessary lenses behind the mirror to correct these errors in the shape of the eyeball.

5. Kühne's Method.—If an artificial eye is not at hand, a very suitable arrangement is that devised by Kühne. Paint a disc to resemble the normal fundus when it is seen with the ophthalmoscope. Remove the eye-piece—long one—from an ordinary microscope. Screw out the lower lens of the eye-piece, fix in the painted disc, and block up the lower aperture with a piece of cork. Fix the eye-piece in a suitable holder, and use it instead of an eye to be examined.

LESSON LXV.

TOUCH—SMELL—TASTE—HEARING.

1. Touch. The Sense of Locality.

(*a.*) Cause a person to shut his eyes, touch some part of his body with a pin, and ask him to indicate the part touched.

(*b.*) Use a small pair of wooden compasses, or an ordinary pair of dividers with their points guarded by a small piece of cork, or Sieveking's Æsthesiometer. Apply lightly the points of the compasses simultaneously to different parts of the body, and ascertain at what distance apart the points are felt as two. The following is the order of sensibility :—Tip of tongue (1·1 mm.), tip of the middle finger (2·3), palm (8 to 9), forehead (22), back of hand (31·6), back (66).

(*c.*) Test as in (*b.*) the skin of the arm, beginning at the shoulder and passing downwards. Observe that the sensibility is greater as one tests towards the fingers, and also in the transverse than in the long axis of the limb.

(*d.*) By means of a spray-producer spray the back of the hand with ether, and observe how the sensibility is abolished.

(*e.*) In all cases compare the results obtained on both sides of the body.

(*f.*) Illusions.   Aristotle's Experiment.—Cross the middle over the index finger, as in Fig. 142, roll a small ball between the fingers; one has a distinct impression of two balls. Or, cross the fingers in the same way, and rub them against the point of the nose. The same illusion is experienced.

2. The Sense of Temperature.

(*a.*) Ask the person experimented on to close his eyes. Use two test-tubes, one filled with cold and the other with hot water; or two spoons, one hot and one cold. Apply one or other to different parts of the surface, and ask the person to say whether the touching body is hot or cold. Test roughly the sensibility of different parts of the body with cold and warm metallic-pointed rods.

Fig. 142.

(*b.*) Touch a ball of fur, wood, and metal. Observe that the metal feels coldest, although all the objects have the same temperature.

(*c.*) Plunge the hand into water at 36° C. One experiences a feeling of heat. Then plunge it into water at 30° C., at first it feels cold, because heat is abstracted from the hand. Plunge the other hand direct into water at 30° C. without previously placing it in water at 36° C., it will feel pleasantly warm.

(*d.*) Hold one hand for a time in water at 10° C., and afterwards place it in water at 20° C., at first the latter causes a sensation of heat, which soon gives place to that of cold.

(*e.*) Test with the finger the acuteness of the sense of temperature—*i.e.*, in two given fluids of different temperatures, what fraction of a degree C. can be distinguished. One can usually distinguish $\frac{2}{5}°$, although the acuteness is greater when the fluids are about 30° C.

(*f.*) Use two brass tubes (5 cm. long and 1 cm. in diam.), terminating in a point. Cover both, all except the point, with india-rubber tubing. Fill one with warm water and the other with cold. Test the position of the **warm and cold points** on another person on various parts of the skin.

3. The Sense of Pressure.

(*a.*) Rest the dorsum of the hand on a table, cover a small area of the palm with a non-conducting material—*e.g.*, a wooden disc or vulcanite plate—and on the latter place different weights, and estimate the smallest difference of weight which can be appreciated.

(*b.*) Dip the hand into water of the same temperature as the hand, or a finger into mercury. The greatest sensation is felt at the plane of the fluid in the form of a ring, but even this is best felt on moving the hand up and down.

4. Law of Peripheral Projection.

(*a.*) Dip the elbow in ice-cold water; at first one feels the sensation of cold, owing to the effect on the cutaneous nerve-endings. Afterwards, when the trunk of the ulnar nerve is affected, the pain is felt in the skin of the ulnar side of the hand where the nerve terminates.

(*b.*) Press the ulnar nerve at the elbow, the prickling feeling is referred to the skin on the ulnar side of the hand.

5. Illusions.

(*a.*) Place a thin disc of *cold* lead the size of a florin on the forehead of a person whose eyes are closed, remove the disc, and on the same spot place two warm discs of equal size. The person will judge the latter to be about the same weight, or lighter, than the single cold disc.

(*b.*) Compare two similar wooden discs, and let the diameter of one be slightly greater than that of the other. Heat the smaller one over 50°C., and it will be judged heavier than the larger cold one.

(*c.*) Lay on different parts of the skin a small square piece of paper with a small central hole in it. Let the person close his eyes, while another person gently touches the uncovered piece of skin with cotton wool, or brings near it a hot body. In each case ask the observed person to distinguish between them. He will always succeed on the volar side of the hand, but occasionally fail on the dorsal surface of the hand, the extensor surface of the arm, and very frequently on the skin of the back.

6. The Muscular Sense.

(*a.*) With the arm and hand unsupported, the eyelids closed, and the same precautions as in 3 (*a.*), determine the smallest difference which can be perceived between two weights. It will be less than in cartridges filled with a known weight of shot, and tested by the pressure sense alone. The cartridges, *e.g.*, 100 grms., are numbered, but they are so made as to have a small increasing increment of weight. They are alike in external appearance.

(*b.*) Take two equal iron or lead weights, heat one and leave the other cold. The cold one will feel the heavier.

7. Taste and Smell.—Prepare a strong solution of sulphate of quinine, with the aid of a little sulphuric acid to dissolve it (*bitter*), a 5 per cent. solution of sugar (*sweet*), a 10 per cent.

solution of common salt (*saline*), and a 1 per cent. solution of acetic acid (*acid*).

(*a.*) Wipe the tongue dry, lay on its tip a crystal of sugar. It will not be tasted until it is dissolved.

(*b.*) Apply a crystal of sugar to the tip and another to the back of the tongue. The sweet taste is more pronounced at the tip. Do the same with the sugar solution, applying it by means of a small camel's-hair brush.

(*c.*) Repeat the same with sulphate of quinine in powder and in liquid. It will scarcely be tasted on the tip of the tongue, but will be tasted immediately on the back part of the dorsum.

(*d.*) Ascertain where saline and acid substances are tasted most acutely.

(*e.*) Connect two zinc terminals with a large Grove's battery, apply them to the upper and under surface of the tongue, and pass a constant current through the tongue. An acid taste will be felt at the positive, and an alkaline one at the negative pole.

(*f.*) Close the nostrils, shut the eyes, and attempt to distinguish by taste alone between an apple and a potato.

8. Hearing.

(*a.*) Hold a ticking watch between your teeth, or touch the upper incisors with a vibrating tuning-fork, close both ears, and observe that the ticking is heard louder. Unstop one ear, and observe that the ticking is heard loudest in the stopped ear.

(*b.*) Hold a vibrating tuning-fork on the incisor teeth until you cannot hear it sounding. Close one or both ears and you will hear it.

(*c.*) Listen to a ticking watch or a tuning-fork kept vibrating electrically. Close the mouth and nostrils, and take either a deep inspiration or deep expiration so as to alter the tension of the air in the tympanum; in both cases the sound is diminished.

(*d.*) Connect two telephones in circuit with a vibrating Neef's hammer of an induction machine, and place a telephone to each ear; one hears the sound as if it came from within one's own head in the vertical median plane.

(*e.*) With a blindfolded person test his sense of the direction of sound—*e.g.*, by clicking two coins together. It is very imperfect. Let a person press both auricles against the side of the head, and hold both hands vertically in front of each meatus. On a person making a sound in front, the observed person will refer it to a position behind him.

(*f.*) Test the highest audible sound by means of Galton's whistle.

INDEX.

www.ingramcontent.com/pod-product-compliance
Lightning Source LLC
Chambersburg PA
CBHW021502210326

41599CB00012B/1104